MARINE MAMMAL PROTECTION ISSUES

DEREK L. CARUANA
EDITOR

Nova Science Publishers, Inc.

New York

Copyright © 2010 by Nova Science Publishers, Inc.

All rights reserved. No part of this book may be reproduced, stored in a retrieval system or transmitted in any form or by any means: electronic, electrostatic, magnetic, tape, mechanical photocopying, recording or otherwise without the written permission of the Publisher.

For permission to use material from this book please contact us:
Telephone 631-231-7269; Fax 631-231-8175
Web Site: http://www.novapublishers.com

NOTICE TO THE READER

The Publisher has taken reasonable care in the preparation of this book, but makes no expressed or implied warranty of any kind and assumes no responsibility for any errors or omissions. No liability is assumed for incidental or consequential damages in connection with or arising out of information contained in this book. The Publisher shall not be liable for any special, consequential, or exemplary damages resulting, in whole or in part, from the readers' use of, or reliance upon, this material. Any parts of this book based on government reports are so indicated and copyright is claimed for those parts to the extent applicable to compilations of such works.

Independent verification should be sought for any data, advice or recommendations contained in this book. In addition, no responsibility is assumed by the publisher for any injury and/or damage to persons or property arising from any methods, products, instructions, ideas or otherwise contained in this publication.

This publication is designed to provide accurate and authoritative information with regard to the subject matter covered herein. It is sold with the clear understanding that the Publisher is not engaged in rendering legal or any other professional services. If legal or any other expert assistance is required, the services of a competent person should be sought. FROM A DECLARATION OF PARTICIPANTS JOINTLY ADOPTED BY A COMMITTEE OF THE AMERICAN BAR ASSOCIATION AND A COMMITTEE OF PUBLISHERS.

LIBRARY OF CONGRESS CATALOGING-IN-PUBLICATION DATA

Marine mammal protection issues / editor, Derek L. Caruana.
    p. cm.
Includes index.
ISBN 978-1-60741-540-4 (softcover)
1. Marine mammals--Conservation--Government policy--United States. 2. United States. Marine Mammal Protection Act of 1972. 3. United States. National Marine Fisheries Service I. Caruana, Derek L. II. Buck, Eugene H. III. United States. Government Accountability Office.
QL713.2.M28 2009
599.5'177--dc22
                    2009039021

Published by Nova Science Publishers, Inc. † New York

FISH, FISHING AND FISHERIES

MARINE MAMMAL PROTECTION ISSUES

FISH, FISHING AND FISHERIES

Additional books in this series can be found on Nova's website at:

https://www.novapublishers.com/catalog/index.php?cPath=23_29&series
p=Fish%2C+Fishing+and+Fisheries

Additional e-books in this series can be found on Nova's website at:

https://www.novapublishers.com/catalog/index.php?cPath=23_29&series
pe=Fish%2C+Fishing+and+Fisheries

CONTENTS

Preface   **vii**

Chapter 1   The Marine Mammal Protection Act: Reauthorization
                 Issues   **1**
                 Eugene H. Buck

Chapter 2   National Marine Fisheries Service: Improved Economic
                 Analysis and Evaluation Strategies Needed for Proposed
                 Changes to Atlantic Large Whale Protection Plan   **61**
                 GAO

Chapter 3   National Marine Fisheries Service: Improvements are
                 Needed in the Federal Process Used to Protect Marine
                 Mammals from Commercial Fishing   **115**
                 GAO

Chapter Sources   **171**

Index   **173**

PREFACE

The National Marine Fisheries Service (NMFS) developed the Atlantic Large Whale Take Reduction (ALWTR) plan to protect endangered large whales from entanglements in commercial fishing gear, which can cause injury or death. Because whales continued to die after the ALWTF plan went into effect, NMFS proposed revisions in 2005. The author of this book discusses these issues, as well as the Marine Mammal Protection Act (MMPA), which requires the NMFS to establish take reductions teams for certain marine mammals to develop measures to reduce their incidental takes. Other bills that specifically address marine mammal regulatory and management issues are examined as well. Furthermore, while some of these issues can be addressed administratively, in regulations proposed and promulgated by the National Marine Fisheries Service, the U.S. Fish and Wildlife Service, or the USDA Animal and Plant Health Inspection Service, others likely would require statutory change. This book lays out the range of issues likely to be raised during reauthorization debate, the reasons behind them, and possible proposals that could be offered to address these concerns. This book consists of public domain documents which have been located, gathered, combined, reformatted, and enhanced with a subject index, selectively edited and bound to provide easy access.

Chapter 1 - The Marine Mammal Protection Act (MMPA) was last reauthorized in 1994. The MMPA's authorization of appropriations expired at the end of FY1999. At issue for Congress are the terms and conditions of provisions designed to reauthorize and amend the MMPA to address a variety of concerns relating to marine mammal management. In the 109th Congress, the House passed a bill to reauthorize and amend the MMPA, but no further action was taken on this measure. The 110th Congress may again consider measures to amend and reauthorize the MMPA as well as bills to address specific marine mammal regulatory and management issues.

Several issues that may arise in reauthorization relate to modifying management of the interactions between marine mammals and commercial fishing operations. Other concerns relate to marine mammals in captivity and subsistence use of marine mammals by Native Americans. Additional issues include providing for trade in marine mammal products, managing robust marine mammal stocks, understanding the effect of noise on marine mammals, fostering international cooperati on, regulating large incidental takes, modifying the scientific research permit process, improving agency compliance with MMPA deadlines, facilitating marine mammal research by federal scientists, dealing with harassment of marine mammals, considering a directed research program, and appropriating adequate funding for federal agency programs. While some of these issues could be addressed administratively, in regulations proposed and promulgated by the National Marine Fisheries Service, the U.S. Fish and Wildlife Service, or the USDA Animal and Plant Health Inspection Service, others likely would require statutory changes.

Most potential participants in the reauthorization debate anticipate extended negotiations on some of these issues. Although the authorization for appropriations expired at the end of FY1999, the MMPA itself did not expire. Eventually, however, an extension of funding authority may need to be considered to continue federal program operations. Most of the issues associated with this law are not timesensitive, and a number of oversight hearings have been held to increase understanding of various issues, positions, and possibilities.

This chapter lays out the range of issues likely to be raised during any reauthorization debate, the reasons behind them, and possible proposals that could be offered to address these concerns

Chapter 2 - The National Marine Fisheries Service (NMFS) developed the Atlantic Large Whale Take Reduction (ALWTR) plan to protect endangered large whales from entanglements in commercial fishing gear, which can cause injury or death. Because whales continued to die after the ALWTR plan went into effect, NMFS proposed revisions in 2005. GAO was asked to review these proposed revisions, including (1) their scientific basis and uncertainties regarding their effectiveness, (2) NMFS's plans to address concerns about the feasibility of implementing them, (3) the extent to which NMFS fully assessed the costs to the fishing industry and impacts on fishing communities, and (4) the extent to which NMFS developed strategies for fully evaluating their effectiveness. GAO reviewed the proposed changes to the ALWTR plan and obtained the views of NMFS officials, industry representatives, scientists, and conservationists.

Chapter 3 - Because marine mammals, such as whales and dolphins, often inhabit waters where commercial fishing occurs, they can become entangled in fishing gear, which may injure or kill them—this is referred to as "incidental take." The 1994 amendments to the Marine Mammal Protection Act (MMPA) require the National Marine Fisheries Service (NMFS) to establish take reduction teams for certain marine mammals to develop measures to reduce their incidental takes. GAO was asked to determine the extent to which NMFS (1) can accurately identify the marine mammal stocks—generally a population of animals of the same species located in a common area—that meet the MMPA's requirements for establishing such teams, (2) has established teams for those stocks that meet the requirements, (3) has met the MMPA's deadlines for the teams subject to them, and (4) evaluates the effectiveness of take reduction regulations. GAO reviewed the MMPA, and NMFS data on marine mammals, and take reduction team documents and obtained the views of NMFS officials, scientists, and take reduction team members.

In: Marine Mammal Protection Issues
Editors: Derek L. Caruana
ISBN: 978-1-60741-540-4
© 2010 Nova Science Publishers, Inc.

Chapter 1

THE MARINE MAMMAL PROTECTION ACT: REAUTHORIZATION ISSUES

Eugene H. Buck

Natural Resources PolicyResources, Science, and Industry Division

SUMMARY

The Marine Mammal Protection Act (MMPA) was last reauthorized in 1994. The MMPA's authorization of appropriations expired at the end of FY1999. At issue for Congress are the terms and conditions of provisions designed to reauthorize and amend the MMPA to address a variety of concerns relating to marine mammal management. In the 109th Congress, the House passed a bill to reauthorize and amend the MMPA, but no further action was taken on this measure. The 110th Congress may again consider measures to amend and reauthorize the MMPA as well as bills to address specific marine mammal regulatory and management issues.

Several issues that may arise in reauthorization relate to modifying management of the interactions between marine mammals and commercial fishing operations. Other concerns relate to marine mammals in captivity and subsistence use of marine mammals by Native Americans. Additional issues include providing for trade in marine mammal products, managing robust marine mammal stocks, understanding the effect of noise on marine mammals, fostering international cooperati on, regulating large incidental takes, modifying the scientific research permit process, improving agency compliance with MMPA deadlines, facilitating marine mammal research by federal scientists, dealing with harassment of marine mammals, considering a directed research program, and appropriating adequate funding for federal agency programs. While some of these issues could be addressed administratively, in regulations proposed and promulgated by the National Marine Fisheries Service, the U.S. Fish and Wildlife Service, or the USDA Animal and Plant Health Inspection Service, others likely would require statutory changes.

Most potential participants in the reauthorization debate anticipate extended negotiations on some of these issues. Although the authorization for appropriations expired at the end of FY1999, the MMPA itself did not expire. Eventually, however, an extension of funding authority may need to be considered to continue federal program operations. Most of the issues associated with this law are not timesensitive, and a number of oversight hearings have been held to increase understanding of various issues, positions, and possibilities.

This chapter lays out the range of issues likely to be raised during any reauthorization debate, the reasons behind them, and possible proposals that could be offered to address these concerns.

INTRODUCTION

The Marine Mammal Protection Act (MMPA) of 1972 (P.L. 92-522, as amended; 16 U.S.C. §§1361, et seq.) was last reauthorized in 1994 by P.L. 103-238. The authorization of appropriations under the MMPA expired at the end of FY1999. The 104th, 105th, 106th, 108th, and 109th Congresses enacted additional amendments addressing single or limited issues (see "Miscellaneous MMPA Amendments"); no MMPA amendments were enacted by the 107th Congress.[1] At issue for Congress are the terms and conditions of provisions to reauthorize and amend the MMPA to address a variety of concerns related to marine mammal management. Legislation introduced, but not enacted, in the 105th, 106th, 107th, 108th, and 109th Congresses suggests a number of issues that may be discussed during a reauthorization debate. To identify a larger universe of potentially relevant concerns, the Congressional Research Service queried commercial fishing, scientific research, public display,[2] animal protection, Native American, and environmental interests to identify issues that might surface during a reauthorization debate.[3] This chapter identifies these concerns and provides background to facilitate a better understanding of various positions on these issues. These concerns, along with other factors, may be considered as Congress determines whether and how to address MMPA reauthorization.

Other than recommendations contained in reports to Congress mandated by the MMPA Amendments of 1994 (discussed later in this chapter) and in testimony presented at a June 29, 1999 oversight hearing before the House Resources Subcommittee on Fisheries Conservation, Wildlife, and Oceans, the Clinton Administration did not release any comprehensive proposals related to MMPA reauthorization. In the 109th Congress, H.R. 4075 incorporated some of the MMPA amendments proposed by the Bush Administration.[4] Congress has been active on marine mammal protection issues in recent years, responding primarily to balancing concerns of the commercial fishing industry and environmental interests. Congress generally views the MMPA as working well, but possibly needing changes to address an increasing number of concerns that have arisen since the 1994 amendments. In the House, the Committee on Natural Resources has jurisdiction over any MMPA reauthorization legislation. In the Senate, the Committee on Commerce, Science, and Transportation has jurisdiction over any legislation on this issue.

CONSTITUENCY GROUPS

An array of groups and individuals hold common and conflicting interests in our nation's marine mammals. Despite their diversity, they generally share the goals of ensuring sustainable marine mammal populations and maintaining healthy marine ecosystems. These groups, however, sometimes disagree about how best to achieve these goals and use these common resources, and thus, conflict is inevitable. As Congress considers reauthorization of the MMPA, these diverse groups will advocate a wide variety of policy proposals.[5]

Commercial Fishing Industry

There were more than 65,000 commercial fishing vessels estimated to be operating in U.S. marine fisheries in 2001,[6] and 3,242 processor and wholesale plants employing 65,690 individuals in 2004.[7] In 2005, the total catch of marine fish was more than 9.6 billion pounds, with an estimated ex-vessel value[8] of more than \$3.9 billion.[9] For 2005, the overall economic contribution of commercial fishing to gross national product (in value added) was estimated to exceed \$32.9 billion.[10]

This sector is chiefly concerned with ensuring sustainable fisheries that balance environmental protection with the continued short-term and long-term viability of the industry. An additional concern is how best to manage conflicts between increasingly abundant marine mammals and commercial fishing. Within this sector is a diverse group of interests, each with specific concerns regarding the rational use of living marine resources and the allocation of resources among user groups. These sectors divide according to scale of operation; type of activity (fishermen, catcher-processor, processor); type of fishing gear used (trawl, longline, gillnet, pots, seine); and location (inshore or offshore).

Environmental Groups

More than 50 U.S. environmental and conservation organizations focus primarily or largely on marine issues. Membership in these groups ranges into the millions. With respect to the MMPA, environmental groups are principally concerned with the lack of assessment data for many managed stocks, the direct and indirect harm to less resilient marine species (including marine mammals), the protection of marine biodiversity, and the continuing loss of marine habitat.

Public Display Community

This community includes about 200 U.S. marine life parks, aquariums, and zoos dedicated to the conservation of marine mammals and their environments through public display, education, and research. The public display community has not taken a formal position on any of the issues raised in this chapter.

Animal Protection Advocates

More than 30 U.S. animal protection organizations have programs focusing on the protection of marine mammals and other marine species. Animal protection groups are concerned with impacts on individual animals as well as species, with harassment as well as killing and injury, and with captive as well as free-ranging animals. They share concerns with environmental groups regarding habitat loss and degradation, but are also concerned with intentional and incidental takes that result in animal suffering. Their key focus is on protection.

Native Americans

Because of their culture, tradition, and subsistence needs, many tribes and indigenous groups are concerned about the management of marine mammals. Some Alaska Native groups are represented by commissions (e.g., Eskimo Walrus Commission, Alaska Eskimo Whaling Commission, Harbor Seal Commission, Aleut Marine Mammal Commission, Alaska Nanuuq Commission, Alaska Sea Otter and Steller Sea Lion Commission) that coordinate management of certain species with federal agencies. The long-term goals of tribes and indigenous groups generally include economic stability, resource sustainability, and regulatory certainty. Of particular concern during MMPA reauthorization will be cooperative management of marine mammals, which they believe fosters economic vitality, environmental health, and rational management of natural resources.

Marine Mammal Scientists

Scientists from academia, the private sector, and state and federal agencies are principally involved in analyzing the ecological, social, and economic effects of MMPA provisions and marine mammal management policy. Like the other groups, they are concerned with the health and integrity of marine ecosystems and the rational use of marine resources. Specifically, they are interested in the availability of adequate funding and accurate data to perform the necessary analyses. Such scientists are also often members of or associated with other constituent groups.

Marine Mammal Managers

Federal and state marine mammal managers are charged with implementing the MMPA and complementary state programs. Because of this responsibility, their interests and concerns are more keenly focused on the pragmatic aspects of the MMPA. Specifically, they are interested in clarity in the intent of management requirements and in federal appropriations to fund data collection and research.

MARINE MAMMAL PROTECTION ACT

Congress enacted the Marine Mammal Protection Act (MMPA) in 1972, due in part to the high level of dolphin mortality in the eastern tropical Pacific tuna fishery (estimated at more than 400,000 animals per year in the late 1960s). The MMPA established a moratorium[11] on the "taking" of marine mammals in U.S. waters and by U.S. nationals on the high seas.[12] The MMPA also established a moratorium on importing marine mammals and marine mammal products into the United States. The MMPA protects marine mammals from "clubbing, mutilation, poisoning, capture in nets, and other human actions that lead to extinction." It also expressly authorized the Secretaries of Commerce and the Interior to issue permits for the "taking" of marine mammals for certain purposes, such as scientific research and public display.

Under the MMPA, the Secretary of Commerce, acting through the National Marine Fisheries Service (NMFS, in the National Oceanic and Atmospheric Administration, also popularly referred to as "NOAA Fisheries"), is responsible for the conservation and management of whales, dolphins, porpoises, seals, and sea lions. The Secretary of the Interior, acting through the U.S. Fish and Wildlife Service (FWS), is responsible for walruses, sea otters, polar bears, manatees, and dugongs. This division of authority derives from agency responsibilities as they existed when the MMPA was enacted. Title II of the MMPA established an independent Marine Mammal Commission (MMC) and its Committee of Scientific Advisors on Marine Mammals to oversee and recommend actions necessary to meet the requirements of the MMPA. Title III authorized the International Dolphin Conservation Program. Title IV authorized the Marine Mammal Health and Stranding Response Program. Title V implemented the Agreement Between the United States and the Russian Federation on the Conservation and Management of the Alaska-Chukotka Polar Bear Population.

Prior to passage of the MMPA, states were responsible for managing marine mammals on lands and in waters under their jurisdiction. The MMPA shifted all marine mammal management authority to the federal government. It provides, however, that management authority, on a species-by-species basis, could be returned to a state that adopts conservation and management programs consistent with the purposes and policies of the MMPA.[13] It also provides that the moratorium on taking can be waived by the federal government or states with management authority for specific purposes, if the taking will not disadvantage the affected species or population. Permits may be issued to take or import any marine mammal species, including depleted species, for scientific research or to enhance the survival or recovery of the species or stock. Non-depleted species may be taken or imported for purposes of public display. The MMPA allows U.S. citizens to apply for and obtain authorization for taking small numbers of mammals incidental to activities other than commercial fishing (e.g., offshore oil and gas exploration and development), if the taking would have a negligible impact on any marine mammal species or stock, and if monitoring requirements and other conditions are met.

The MMPA's moratorium on taking does not apply to any resident Alaskan Indian, Aleut, or Eskimo who dwells on the coast of the North Pacific (including the Bering Sea) or Arctic Oceans (including the Chukchi and Beaufort Seas), if such taking is for subsistence purposes[14] or for creating and selling authentic Native articles of handicrafts and clothing, and is not done wastefully. However, such taking can be regulated or even prohibited if the Secretary determines a stock is depleted. The MMPA also provides for co-management of marine mammal subsistence use by Alaska Native groups, under which authority Native commissions have been established.

The MMPA also authorizes the taking of marine mammals incidental to commercial fishing operations. In 1988, most U.S. commercial fish harvesters were exempted from otherwise applicable regulations and permit requirements for five years, pending development of an improved system to govern the incidental taking of marine mammals in the course of commercial fishing operations.[15] The taking of marine mammals incidental to the eastern tropical Pacific tuna fishery is governed by specific and separate provisions in Title III of the MMPA.

The Endangered Species Act of 1973 (ESA; P.L. 93-205, asamended; 16 U.S.C. §§1531, et seq.) provides additional protection for some marine mammal species that have been determined to be threatened or endangered with extinction. When protective actions are taken under both ESA and MMPA authorities, interactions between implementation efforts under these two statutes may increase management complexity and legal uncertainty in dealing with some species, such as the southern sea otter in California.

1994 MMPA REAUTHORIZATION

The 1988 commercial fishing exemption expired at the end of FY1993, and new provisions were enacted in P.L. 103-238, which reauthorized the MMPA through FY1999.[16] These new provisions indefinitely authorized the taking of marine mammals incidental to commercial fishingoperations and provided for (1) preparing assessments for all marine mammal stocks in waters under U.S. jurisdiction, (2) developing and implementing Take Reduction Plans for stocks that may be reduced or are being maintained below their optimum sustainable population levels due to interactions with commercial fisheries, and (3) studying pinniped[17]-fishery interactions. In addition, the 1994 amendments substantially changed provisions relating to public display of marine mammals, authorized imports of polar bear trophies from Canada, authorized the limited lethal removal of pinnipeds, and enacted a general authorization for research involving only low levels of harassment.

Implementation of the 1994 Amendments

Implementation of the 1994 MMPA amendments by NMFS and FWS has been controversial on several issues. In some cases, implementation of new provisions took more than the full five years of the authorization to complete. One of the most difficult and controversial amendments to implement was the convening of Take Reduction Teams (TRTs)[18] and the development of Take Reduction Plans by these TRTs. Critics believe an insufficient number of TRTs have been convened and that development of plans is far behind schedule.

NMFS has convened nine TRTs to reduce bycatch of strategic stocks of marine mammals in selected commercial fisheries. One of these, the Atlantic Offshore Cetacean TRT, was disbanded in August 2001 due to changes in the affected fisheries. Another, the Mid-Atlantic TRT, became the Mid-Atlantic

Harbor Porpoise TRT, because of priority given to particularly vulnerable harbor porpoise. Six of the remaining TRTs address bycatch issues on the Atlantic Coast, while the remaining TRT addresses marine mammal bycatch in the Pacific driftnet fishery for swordfish and sharks.[19]

Several TRTs have yet to be convened and plans developed. Although NMFS recognizes that fishery-related mortality exceeds the PBR level in some marine mammal stocks, no new TRTs are to be convened until additional funds are appropriated or redirected from existing Take Reduction Plans that have been declared successful. Congress recognized that funds would be limited and established criteria for prioritization of this effort in 16 U.S.C. §1387(f)(3).

Overall, NMFS believes that the time allowed by the MMPA to convene a TRT and develop a plan has been adequate. However, NMFS found it difficult to publish a final rule based on a plan in the time allotted by the MMPA, due primarily to the complexity and difficulty of implementing regulations that minimize impacts to the industry as required by the MMPA, and by the economic analyses and requirements of other statutes. The difficulties in meeting statutory deadlines and implementing plans for these strategic stocks has been both frustrating to many, sometimes resulting in litigation, and satisfying to others in that serious bycatch/fishery issues have been addressed.

In response to the 1994 MMPA amendments at 16 U.S.C. §1386, NMFS and FWS have completed reports assessing more than 170 different marine mammal stocks as required by the MMPA,[20] and have developed a list of fisheries that monitor their annual takes of marine mammals by stock. These lists are frequently being revised and are also the target of controversy as new information is incorporated into the assessments. However, some of the stock assessments conducted by FWS have been criticized for using outdated (e.g., decades old) data. In addition, Alaska Native interests continue to be concerned that some stock assessment reports have little information on incidental take from commercial fishing operations. The sparse information in these reports, based on data collected during the 1988-1992 exemption, was the result of an emphasis on some U.S. fisheries having interactions with marine mammals at a rate that was considered more serious than that occurring in fisheries of concern to Native Alaskans. An observer program was therefore not initiated in Alaska until 1998. NMFS anticipates becoming better able to address the concerns of Alaska Natives.

The 1994 amendments also directed the federal government to undertake an ecosystem-based research and monitoring program for the Bering Sea to identify the causes of ecosystem decline. There is controversy over the extent to which this provision has been fulfilled. Meanwhile, the Alaska Native community would like to initiate a complimentary effort to understand Bering Sea ecological processes by drawing upon traditional Native knowledge and wisdom. The Alaska Native community was unable to obtain public funding to convene meetings amongaffected villages to review the draft federal Bering Sea research plan, and eventually sought independent funding to support a March 1999 Bering Sea conference.

Miscellaneous MMPA Amendments

Subsequent to the 1994 reauthorization, several additional MMPA amendments were enacted separately:

- Section 405(b)(3) of P.L. 104-297 amended the MMPA's definition of the term "waters under the jurisdiction of the United States."
- In P.L. 105-18, §2003 provided a "good samaritan" exemption allowing individuals to free marine mammals entangled in fishing gear or debris, while §5004 modified the requirements for the importation of polar bear parts from polar bears legally harvested in Canada before the MMPA Amendments of 1994 were enacted.
- P.L. 105-42 modified dolphin conservation provisions of the MMPA applicable to the eastern tropical Pacific tuna seine fishery and specified under what conditions tuna products can be labeled "dolphin-safe."
- Administrative provisions for the U.S. Fish and Wildlife Service in P.L. 105-277 clarified that polar bear trophy permit fees remain available until expended for cooperative research and management programs.
- Title II of P.L. 106-555 authorized grants to benefit marine mammal stranding programs. !Section 149 of P.L. 108-108 permitted the importation of polar bears from Canada harvested prior to the enactment of final regulations.
- Section 319 of P.L. 108-136 modified the MMPA's definition of harassment and provisions relating to taking marine mammals as they relate to military readiness activities and federal scientific research.[21]
- Title IX of P.L. 109-479 implemented the Agreement Between the United States and the Russian Federation on the Conservation and Management of the Alaska-Chukotka Polar Bear Population.

In compliance with the International Dolphin Conservation Program Act (P.L. 105-42), the Secretary of Commerce, on April 29, 1999, made an initial finding that there was insufficient evidence of significant adverse impact from chase and encirclement of dolphins during tuna fishing.[22] Subsequently, NMFS promulgated a new standard for dolphin-safe tuna in January 2000.[23] However, this standard was challenged byenvironmental groups and overturned by the U.S. District Court for the Northern District of California on April 11, 2000.[24] Although the Department of Commerce appealed this ruling, the 9th Circuit Court of Appeals affirmed the lower court decision in July 2001.[25]

ISSUES FOR CONGRESS

The remainder of this chapter reviews issues that may be raised during discussions on reauthorizing the MMPA. The major issue categories include commercial fishing interactions with marine mammals, marine mammals in captivity, Native Americans and marine mammals, permits and authorizations, and program management and administration. Some of these issues could be addressed administratively, in regulations implemented by NMFS, FWS, or the Animal and Plant Health Inspection Service (APHIS, Department of Agriculture). Others would require legislative action.

Commercial Fishing Interactions with Marine Mammals

Optimum sustainable population

Optimum sustainable population (OSP) is defined in 16 U.S.C. §1362(9) as "the number of animals which will result in the maximum productivity of the population or the species, keeping in mind the carrying capacity of the habitat and the health of the ecosystem of which they form a constituent element." However, the variable nature of populations in marine ecosystems makes it nearly impossible to determine carrying capacity. In addition, for many species, the limiting habitat factors that govern carrying capacity are not known or well understood.[26]

Animal protection, scientific, and environmental interests generally agree that OSP is an important concept for assessing the viability of a population or stock. Some scientists, however, express concern that, if current rather than historic population data are used to calculate OSP, OSP levels may be calculated too low for some marine mammal stocks.[27] Some in the commercial fishing industry, however, argue that OSP, as currently defined, is complex and vague in concept. They contend that "maximum productivity" is difficult to determine and imprecise,[28] and complicates work on developing Take Reduction Plans for marine mammal stocks. The difficulties in declaring that a species is at OSP (1) frustrate fishing industry interests by impeding the ability of the federal government to transfer management authority to states and (2) prevent or delay fishermen from gaining authority to deliberately kill marine mammals. In addition, commercial fishing interests chafe when the MMPA, through OSP, grants marine mammals priority access to certain fish stocks and allocates marine mammals de facto "harvest quotas" in direct competition with and to the detriment of the fishing industry. Environmental and scientific interests counter that marine mammals are part of the marine ecosystem and should have their prey species protected from excessive fishing. These interests also believe that critics within the commercial fishing industry may be too quick to blame marine mammals for reductions in target fish populations where predator-prey relationships are incompletely understood.

Commercial fishing interests would like to see the MMPA amended to modify, simplify, and clarify the definition of OSP as the objective for marine mammal management. Scientific, animal protection, and environmental interests believe that OSP, as the central "core" innovation of the MMPA, should be retained and improved. Other suggestions include directing the MMC to host a workshop, involving marine ecologists, oceanographers, and climatologists, to further examine the methods for determining OSP and its derivative potential biological removal (see section below). Appropriation of funds necessary for this task would probably be required.

Calculating potential biological removal

The potential biological removal (PBR) level is used to establish limits on incidental marine mammal mortality for commercial fishing operations. It is defined in 16 U.S.C. §1362(20) as "the maximum number of animals, not including natural mortalities, that may be removed froma marine mammalstock while allowing that stock to reach or maintain its optimum sustainable population." PBR is calculated by multiplying a stock's minimum population estimate by half the known or presumed maximum net productivity of the stock. This product is multiplied by a fractional multiplier known as the recovery factor.[29] Take Reduction Plans are based on two assumptions: (1) that a stock or population currently within its OSP range will remain so, and (2) that any stock or population below its maximum net productivity level will increase to that level if the total human-caused mortality is kept below the PBR level. However, some scientists believe that both these assumptions might be questionable in light of today's much better information.

MMPA critics in the fishing industry and Native Alaskan community believe that NMFS has been so restrictive in calculating PBRs that the economic viability of certain fisheries (e.g., the New England and mid-Atlantic gillnet fisheries, Bering Sea pollock fishery) is being compromised. NMFS and FWS managers counter that the lack of critical data used in PBR calculations limits their ability to calculate precise PBR values for many species. These issues are particularly acute for Alaskan species where population surveys, productivity rates, and harvest data are absent or based on crude estimates several decades old. Some scientific and animal protection interests, however, are concerned that, if the method for calculating them is changed, PBRs could be set too high to provide adequate incentive for commercial fishermen to develop better ways of targeting and catching certain species of fish (e.g., phasing out indiscriminate harvesting methods).

Segments of the commercial fishing industry would like to have the concept or definition of PBR revised to be less restrictive by, for example, manipulating one of the multipliers (particularly the recovery factor). Scientific, animal protection, and environmental interests believe that PBR is an extremely important concept and an excellent management tool that should be maintained.[30] Without a way to calculate concrete limits on take, they argue, NMFS would have no way of adequately determining the impact of human-caused mortality on marine mammal stocks or of adequately enforcing regulations. Some scientists contend that the necessary monitoring and research to accurately calculate useful PBRs is lacking. These critics suggest that a deadline be set for completingdevelopment of models to address these concerns.

As mentioned in the previous section, suggestions for MMPA reauthorization include directing the MMC to host a workshop, involving marine ecologists, oceanographers, and climatologists, to further examine the methods for determining OSP and its derivative PBR. Considerations for such a workshop might include (1) multiple mortality factors such as subsistence harvest, commercial fishery interactions (including entanglement in net discards), and industrial activities (e.g., noise, contaminants); (2) standardized guidelines for using the recovery factor (e.g., endangered species that continue to decline should use 0.1 or less; endangered but increasing should use 0.2); and (3) variability in natural mortality due to extreme events (e.g., mass stranding, *El Niño*). Appropriation of funds necessary for this task would probably be required.

Zero mortality rate goal

In 16 U.S.C. §1387(b)(3), the MMPA requires "the immediate goal that incidental kill or incidental serious injury of marine mammals permitted in the course of commercial fishing operations be reduced to insignificant levels approaching a zero mortality and serious injury rate within 7 years after April 30, 1994." In July 2004, NMFS defined insignificant levels approaching the ZMRG as 10% or less of the PBR for any stock.[31]

The animal protection and environmental communities believe the objective of approaching the ZMRG must be maintained. However, while marine mammal mortality in many fisheries has been reduced (in some cases, substantially), animal protection and environmental interests do not consider these reductions to be significant. They believe that the ZMRG can be implemented in ways that do not impose burdensome costs on the fishing industry, and that promote marine ecosystem sustainability that is in the interest of all parties. Similar to their reasoning on PBRs, they believe ZMRG must be maintained as a means of encouraging the development and use of more risk-averse fishing methods.

The fishing industry is concerned that ZMRG be implemented in a manner that recognizes a reasonable balance between marine mammal protection and economically viable fisheries, and that can be seen as having been already achieved in many instances. Animal protection and environmental communities generally are supportive of the NMFS definition of approaching the ZMRG — that is, 10% of the PBR or less.

Stock assessment process

The MMPA outlines a stock assessment process in 16 U.S.C. §1386. Several scientists and managers contend that this process is insufficient to assess most marine mammal populations with a reasonable degree of certainty. In addition, these critics as well as various advocacy groups believe that federal funding is insufficient to improve species-specific methods for assessing marine mammal stocks,[32] and that Congress should authorize specific and substantial multi-year funding to improve our basic knowledge of marine mammal populations, especially for Arctic species.[33] Alaskan Native interests suggest that the MMPA (16 U.S.C. §1386(d)) be amended to confer greater authority to Regional Scientific Review Groups, authorizing these groups to exercise more power in addressing concerns of where research is needed, rather than be only advisory. In addition, they suggest amendments to 16 U.S.C. §1386(c) to alter the timing of stock assessment reviews, feeling that healthy stocks may not need review every three years — every five years would be more reasonable. They believe that three years may be too short an interval to detect meaningful trends and can be burdensome on the agency performing the assessments. For most strategic stocks, since little new information is gathered to necessitate an annual review, they believe an assessment every two years might be sufficient.

Deterrence

In 16 U.S.C. §1371(a)(4), the MMPA allows the use of deterrents to discourage marine mammals from damaging fish catch or gear. Currently, the burden falls on the federal government to prove that a deterrent is harmful before it can be prohibited. For example, the long-term effects on marine mammals of acoustical harassment devices (AHDs), such as "seal bombs" and "seal scarers," are not known.[34] NMFS and the Marine Mammal Commission sponsored a 1996 scientific workshop that raised significant concerns about AHDs and recommended that their use be severely limited. NMFS proposed[35] but never finalized deterrence regulations because of the difficulty in identifying measures for safely deterring endangered and threatened marine mammals,[36] and the use of AHDs has increased substantially in recent years. Similarly, some scientists are concerned that there has been insufficient research[37] to determine at what threshold a deterrent device might become harmful to marine mammals.[38]

With huge gaps of knowledge in marine mammal science,[39] some animal protection advocates argue that it would be prudent to allow only proven harmless deterrents for use on marine mammals interacting with fishing vessels and/or fish farms. Some have argued for reliance on the precautionary principle that would require manufacturers to prove that a deterrent does not cause permanent harm to any age/sex class of affected marine mammal species before allowing its use.[40] In addition, some scientists and managers believe that not enough emphasis has been placed on encouraging fishermen to change their fishing practices, rather than use a proven deterrent, to reduce interactions with marine mammals.[41] However, fishermen are likely to make their choice between deterrents and changes in fishing practice on the basis of their relative cost.

Some parties critical of the current situation may endorse proposals to alter the burden of proof for deterrents found in 16 U.S.C. §1371(a)(4)(C); others may support efforts to direct NMFS to study the causes of fishery-marine mammal interaction problems to develop a different basis for regulating deterrents. Others suggest that the MMPA be revised to require permits for AHD users, allowing NMFS to better monitor the amount of ocean noise generated by these devices.[42] NMFS has recommended that Congress consider (1) removing impediments to testing non-lethal deterrent technologies and (2) funding additional research, development, and evaluation of innovative non-lethal pinniped deterrence techniques.[43] Some managers and scientists as well as certain interest groups caution, however, that considerable care must be taken to fully assess the "side effects" of noise and other emissions of non-lethal deterrents to identify any potential for damage to targeted and non-targeted marine mammals, fish that may be more sensitive to noise (e.g., herring, cod, other schooling fish), and divers. Any potential for damage will need to be weighed against the benefits of these deterrents before their use becomes even more widespread.

Reinstate limited authority for intentional lethal taking

Prior to the 1994 MMPA amendments, commercial fishermen were allowed to kill certain pinnipeds as a last resort to protect their gear and catch. The 1994 amendments eliminated authorization for such lethal taking and replaced it with authority to use deterrence measures that do not kill or seriously injure marine mammals. However, conflicts between fishermen and pinnipeds have become more frequent, and economic losses have increased. NMFS has recommended that Congress consider authorizing the intentional lethal taking of California sea lions and Pacific harbor seals in specific areas and fisheries to protect gear and catch until effective non-lethal methods are developed.[44] Critics oppose reinstating this authority, fearing that allowing fishermen to kill California sea lions and Pacific harbor seals could reduce the incentive to modify fishing practices or develop non-lethal deterrents, and would likely result in accidental kills of similar-appearing species that are endangered, such as the ESA-listed Steller sea lion.[45] These critics suggest that more attention be given to modifying fishing practices and fishery management policies to reduce contact between commercial fishermen and marine mammals. One possible means for accomplishing this might involve the creation of marine protected areas that encompass key marine mammal habitats.[46] In particular, animal protection advocates strongly oppose any reinstatement of intentional lethal taking, fearing the increased risks of merely injuring animals and causing significant suffering as shown by the number of live-stranded sea lions that are sent to rehabilitation centers after having been illegally shot.

With regard to sea otters rather than pinnipeds, Washington State sea urchin fishermen are becoming more concerned about harmful interactions by increasingly abundant sea otters, and may seek some means for limiting or controlling sea otter abundance to benefit the sea urchin fishery. The state lists sea otters as endangered, but no federal protection is afforded this population under the ESA. However, a 1996 stock assessment report prepared under MMPA authority indicated this population was below OSP. In addition, Alaskans who blame sea otters, in part, for declining fish catch may advocate a more liberal killing of sea otters by Alaska Natives interested in expanding commercial tradein handicrafts made from their fur. Others, however, are concerned about reported recent declines in Alaska sea otter abundance. Animal protection groups rigorously oppose proposals to lethally take sea otters.

Integration with fishery management

On several issues, observers suggest that better integration between the management programs under the MMPA and the Magnuson-Stevens Fishery Conservation and Management Act might be helpful.[47] Currently, no formal mechanism exists for interaction between Take Reduction Teams (TRTs) and the regional fishery management council committees, established under the Magnuson-Stevens Act, which assess fish stocks, determine total allowable catch (TAC), and make other fishery management decisions. However, marine mammal take reduction is clearly an essential part of reducing fishery bycatch and other incidental mortalities associated with fisheries. The Steller Sea Lion Recovery Team has so far been the only quasi-TRT that has been included in formulating fishery management plans (i.e., by the North Pacific Fishery Management Council for Gulf of Alaska groundfish and for Bering Sea/Aleutian Islands groundfish). Some marine mammal scientists suggest amending the MMPA and the Magnuson-Stevens Act to require TRT input in fishery management planning, to better address marine mammal-fisheries interaction problems.

Fishery impacts and southern sea otters

Because vessels conducting trap and other inshore fisheries off southern California are often too small to carry observers, monitoring the impacts of these fisheries on southern sea otters has been especially challenging. Without evidence that significant mortality results from these particular fishing activities, funds provided to NMFS under the MMPA are not available to identify and monitor potential sources of mortality for southern sea otters, much less to evaluate how trap design might affect sea otter entrapment or otherwise help identify means to minimize conflicts. Some scientists and managers suggest amending the MMPA to facilitate monitoring in small vessel fisheries and to authorize funding to address potential interactions.

Southern sea otters appear to be attempting to extend their range southward. Such behavior may be significant to the long-term survival of this population, scientists contend. However, the commercial fishing industry opposes any expansion of the southern sea otter's range. When FWS was authorized to establish an experimental population of southern sea otters at San Nicolas Island, one of Southern California's Channel Islands, in 1986, the agency was required to limit the potential impacts of translocated southern sea otters on existing commercial fisheries and remove sea otters from a management zone south of Point Conception.[48] In late 2005, FWS proposed that this translocation program be terminated.[49] Commercial fishermen suggest that the MMPA and the ESA might be amended to impose more stringent requirements on managing populations to limit their potential to conflict with existing uses. Opposing this, some environmental and animal protection interests suggest that language establishing the 1986 experimental population and translocation be repealed, eliminating the management zone and allowing sea otters to expand their range naturally to meet their recovery needs.

Marine Mammals in Captivity

While some issues involving marine mammals in captivity discussed in this section may require amendment of the MMPA, many of these issues could also be addressed under the authority of the Animal Welfare Act (AWA) or be addressed administratively in regulations implemented byAPHIS (Department of Agriculture). Procedurally, Congress faces the decision on whether to treat these issues within the MMPA reauthorization process, to treat them as AWA issues and consider them concurrently with MMPA reauthorization, or to address these issues as strictly AWA concerns to be considered at another time. Congressional oversight of agency implementation of the MMPA and the AWA in some of these issue areas may identify regulatory concerns where further direction from Congress may be helpful in refocusing federal agency implementation of existing law.[50]

Many of the issues in this section reflect the contentious relationship between animal protection interests and holders of captive marine mammals.

These constituencies often disagree on whether, and if so under what conditions, marine mammals should be held in captivity.[51]

Authority for captive marine mammals

Prior to the 1994 MMPA amendments, NMFS, FWS, and APHIS shared responsibility for the care and maintenance of marine mammals held by public display facilities. However, the 1994 MMPA amendments delegated primary authority for captive marine mammals to APHIS for regulation under provisions of the Animal Welfare Act.[52] APHIS conducted a negotiated rulemaking process to revise requirements for the humane handling, care, treatment, and transport of marine mammals in captivity.[53] It involved representatives of animal protection groups, marine mammal facilities, veterinary professionals, trainers, and government managers working cooperatively.

The animal protection community, believing that APHIS's expertise and experience is primarily with non-aquatic species, may propose to return jurisdiction to NMFS and FWS, which they feel are better qualified to monitor marine mammal care and maintenance.[54] On the other hand, some in the public display community see no basis for stripping APHIS of primary authority for captive marine mammals, since they contend that APHIS has a long history of developing and enforcing standards of animal health and care and has vigorously exercised its jurisdiction.[55] This has included conducting broad rulemaking proceedings on revised requirements for marine mammals in captivity. In contrast, the public display community views NMFS and FWS as not typically dealing with or being involved in the animal husbandry sector and having limited expertise in the captive maintenance and care of marine mammals.[56] Critics further assert that giving NMFS and FWS jurisdiction in this area would necessitatean expensive program duplicating what APHIS already administers. Some in the public display community further assert that the majority of problems concerning the quality of care provided captive marine mammals occurred prior to the 1994 MMPA amendments and in privately operated facilities that were not regulated, rather than in regulated public display facilities.

Regardless of who regulates these facilities, some marine mammal scientists and animal protection advocates believe that regulations needto be brought more closely into accord with the physical, psychological, and social needs of marine mammals. In addition, they suggest that existing regulations need to be enforced with more rigor and with less influence from the facilities being regulated.[57] They argue that reliance on the public display community to be forthcoming when explaining the application of particular husbandry practices may be open to question, particularly when public display facilities fear that proprietary interest related to husbandry techniques (e.g., successful captive breeding techniques) might be revealed to competitors. They suggest that Congress consider ways in which successful husbandry techniques might be made more openly available in the interest of benefitting the care of marine mammals throughout the public display industry. Under such conditions, husbandry practices might be standardized to better protect animals.

Export of captive animals

The 1994 MMPA amendments repealed export permit and public notification requirements, replacing them with a 15-day federal agency notification requirement prior to export.[58] NMFS has interpreted export provisions as requiring a letter of comity[59] from the foreign government certifying that the standards of the MMPA are upheld in foreign facilities. In addition, NMFS requires a letter of comity for any further transfer of a marine mammal of U.S. origin by one foreign nation to another foreign nation. Animal protection advocates claim that the current status of some of the marine mammals (dolphins, in particular) shipped from the United States to Honduras, China, Portugal, Tahiti, and other countries since the 1994 repeal is not known. The animal protection community is concerned and may seek to amend the MMPA to restore the export requirements to their original condition (i.e., requiring a permit, with a public comment period as part of the process).[60] Some scientists agree that a requirement for export permits should be reinstated,[61] but believe that MMC and NMFS/FWS review of export permits might make public comment unnecessary. Some suggest that Congress require a full accounting from NMFS/FWS for all past exported marine mammals before allowing any further U.S. animals to be exported. In addition, they suggest the MMPA be amended to require a $25,000 surety bond or insurance policy per exported marine mammal to cover emergency medical and transfer costs in the event of financial or natural disaster at a foreign facility. Animal protection advocates recommend mandatory on-site inspections of foreign facilities before any U.S. marine mammal can be exported. Elements of the public display community, however, believe the current process includes extensive safeguards,[62] and that the prior law requirements were outmoded and cumbersome. Some in the public display community may suggest further amending the MMPA to eliminate the 15-day prior agency notification requirement for exports.[63]

Import of captive animals

Some in the public display community may seek to amend the MMPA to treat the import of marine mammals the same way exports are treated (i.e., agency notification required but no permit required and no public comment solicited). They argue that the current process is cumbersome and unnecessary. The animal protection community would likely oppose such an amendment, desiring to retain and possibly strengthen federal agency review of imports as well as the option for public comment. They believe that a public process with agency review would better protect marine mammals, discouraging the import of certain marine mammals such as those captured specifically for the importing facility.

Scientific research on captive marine mammals

Research on captive marine mammals has provided critical information and a substantial body of literature on many aspects of marine mammal biology.[64] Some scientists assert that research on captive marinemammals maybemoreuseful for certain disciplines (e.g., physiology, immunology, nutrition, hearing sensitivity, and cognitive and acoustic abilities) than others (e.g., acoustic behavior and intra- and inter-species interactions).[65] Some scientists have proposed that more research be conducted on how human activities might affect marine animals.[66] They further contend that research on and observation of marine mammals in captivity affords scientists the opportunity to conduct studies with live animals that are not always possible or practical to do in the wild, and contributes valuable data useful in determining management criteria for wild populations. Many of these scientists believe the MMPA has placed an unreasonable burden on scientific research (e.g., invasive research is seriously impeded).

Other scientists as well as parts of the animal protection community question how much of the research conducted on captive marine mammals actually benefits marine mammals in the wild. These critics may suggest that the MMPA be amended to require that more attention to benefits be given by federal agencies that review permits for scientific research on captive marine mammals. Other scientists are likely to oppose any amendment that might increase their regulatory burden or curtail access to potential research animals. As an alternative to greater restrictions, some scientists suggest amending the MMPA to impose a research requirement on all regulated facilities holding marine mammals, with mandatory peer review of these research programs to ensure that the capture and holding of marine mammals for research is justified.[67]

More extensive medical exams for transferred animals

Although both APHIS and FWS require a health certificate from a licensed veterinarian prior to transporting a marine mammal, the United States does not require any blood tests be made on marine mammals destined for export. In addition, neither NMFS nor FWS requires an exporter to prove that an animal harbors no infections,[68] even if the animal may have been exposed to *Morbillivirus* — a highly contagious, distemper-like disease harmful to some marine mammal species.[69] Therefore, critics assert that some disease-carrying marine mammals could be exported to countries where they might infect marine mammals in that region. Animal protection advocates suggest that the MMPA or the AWA may need to be amended to require more safeguards against transferring pathogens (including antibiotic-resistant pathogens) (a) among captive populations when animals are moved; and (b) to wild populations when captive animals are moved to sea-pens[70] or when a public display facility discharges untreated effluents into the marine environment.[71] More extreme scientific critics suggest that transferred animals should be prohibited from ever being placed in a sea pen or other open enclosure, and that imported and exported marine mammals should be treated like parrots and other exotic birds, with quarantines and thorough medical examinations required at each end of the transfer.

Individuals at some public display facilities believe that these matters have been addressed sufficiently in regulations finalized by APHIS.[72] In addition, an individual associated with the public display community relates that medical examinations prior to transporting marine mammals, regardless of their destination, have been a long-standing practice for many zoos and aquaria. Under such practice and before an animal is transferred, a veterinarian conducts an examination and certifies the animal's healthy condition.[73] They further state that, since humans are not required to be proven disease-free before traveling, it would be ridiculous to impose a higher standard for marine mammals. Managers of public display facilities are exceedingly hesitant to accept any animal that could pose a potential pathogenic threat because of their interest in their investment and the difficulty in replacing animals that die. Furthermore, they assert that there is no documented case where release of a captive marine mammal to the wild or to an open ocean pen, or discharge of facility effluent has contributed to an epidemiological episode in the wild.[74]

Necropsies

Currently necropsies on dead marine mammals are performed in-house by public display facility veterinarians.[75] Prior to the 1994 MMPA amendments, necropsy reports were required to be submitted to NMFS and FWS. Current APHIS standards require such reports to be completed and kept on file at the public display facility for three years.[76] Such medical records are available to APHIS inspectors on-site when requested, but are not submitted to, nor kept on file at, APHIS or any centralized point. The only requirement under the MMPA is to report to NMFS and FWS the "date of death of the marine mammal and the cause of death when determined."[77] Thus, necropsies, which formerly were available to the public under the Freedom of Information Act, are no longer public records.[78]

Animal protection advocates believe that public access to necropsy information is important to protecting the well-being of marine mammals in captivity, and they object to the 1994 changes in necropsy policy. They also fear that captive holding facilities minimize the impact of animal deaths by under-reporting findings of a necropsy, performing an inadequate necropsy, or failing to report actual findings.[79] These critics would like to see the MMPA amended to again require that necropsy reports, in standardized format, be submitted to a federal agency, thus guaranteeing public access to them. In addition, animal protection interests may propose a requirement that necropsies be performed by independent/impartial veterinarians (federally employed, appointed, or contracted veterinarians) and that institutions experiencing a marine mammal death report to APHIS within 48 hours, upon which an official examiner would be dispatched to perform the necropsy or review the tissue samples and examine the carcass. Scientists, however, point out that the more time that passes between death and necropsy, the less there is to learn from the necropsy. Thus, this suggests that, in addition to raising costs, the logistics of implementing an external review also may frustrate the ability to gain worthwhile information. Some scientists suggest an alternative approach that would direct veterinarians employed by the public display facilities to conduct necropsies, but allow veterinarians representing animal protection groups to have access to replicate tissue samples from necropsies, if requested. Critics of current policy may also propose that the MMPA be amended to require submission of necropsies on all animals transferred or exported under MMPA authority. In addition, some critics suggest that APHIS be required to conduct more intensive inspections of facilities holding captive marine mammals whenever mortalities at such facilities exceed a certain annual minimum, such as the deaths of either 2 adult animals or 1 juvenile.

Managers of captive holding facilities state that they ensure good healthcare for their animals by providing licensed veterinary care, thus also protecting themselves from liability and claims of negligence.[80] They assert that there is no evidence that such care is suspect. Furthermore, they point out that necropsies were the subject of a 2001 APHIS rulemaking;[81] because these rules are still being implemented, the need for legislation is unclear for now. If more expensive necropsies were required, the issue of who would pay for them is likely to be controversial. Animal protection interests believe that captive holding facilities should pay for supervised necropsies as part of the costs of captive care; managers of captive animals contend that the federal government should bear the costs if additional outside veterinary services were required.

Genetic mixing

Some federal managers have criticized release programs for captive animals on genetic-mixing grounds. Similar concerns have not been stated about husbandry practices related to the movement of animals between captive facilities. The U.S. Navy's use of Atlantic bottlenose dolphins in open-ocean training exercises in the Pacific where they occasionally integrate with local populations of wild Pacific bottlenose dolphins also has been criticized. Scientists, animal protection advocates, and environmentalists question whether it is responsible management to mix animals originating from different oceans, especially if there is the possibility that they or their offspring might be inadvertently or intentionally released into a wild breeding population. These interests suggest that the MMPA should be amended to address the genetic mixing that invariably occurs when captive animals are moved from one facility to another. MMPA provisions requiring attention to this concern might engender greater confidence if such captive animals later became candidates for release programs.[82] An opposing view encourages genetic mixing within captive populations, especially for species with small populations, as an appropriate husbandry practice to maintain genetic diversity, counter inbreeding within the captive population, and reduce the demand for acquiring new animals from the wild.[83] Some scientists believe that the incidental mixing of captive animals with wild stocks is rare and likely insignificant from an evolutionary perspective. However, they suggest that additional research may be required on these issues before appropriate policy can be determined, recommending a government workshop be convened on the topic.[84]

Wild versus captive survivorship

Claims differ on whether marine mammals live longer, similar, or shorter lifespans in captivity compared to the same species in the wild.[85] Animal protection advocates suggest that the MMPA be amended to direct and fund a government workshop to review the status of knowledge on survivorship in captive and wild marine mammal populations.[86] Such a workshop might determine what, if any, concerns are relevant to the performance of facilities holding such animals and influence the development of appropriate captive care and maintenance standards. Since only a few wild populations are reported to have been studied well enough to provide confident data on survivorship, such a workshop likely would identify additional areas for research on wild populations to obtain data necessary for comparison.

Air quality and noise at facilities

Based on speculation from human studies as well as limited reactivity research on wild cetaceans, local environmental conditions may cause stress in individual animals. Some animal protection advocates suggest that the MMPA be amended to mandate a study of the effect of the local environment (e.g., urban noise, vibrations, air pollution) on animals at captive holding facilities, to identify and substantiate any effect on their life expectancy and general health. Such a study might define abusive levels and help determine appropriate captive care and maintenance standards. Some in the public display community, however, suggest that this concern be addressed administratively, and observe that some aspects already were the subject of APHIS rulemaking.[87] Procedures for monitoring environmental effects on marine mammals also have been incorporated in American Zoo and Aquarium Association guidelines and facility operations manuals.

Rehabilitation and release

Closures of at least 21 North American marine parks since 1990, a diminishing emphasis on marine mammal exhibits in remaining parks, reductions in the military use of marine mammals, and increasingly successful captive breeding programs have led to a surplus of marine mammals in captivity. Because of this surplus, interest has increased concerning the rehabilitation and release to the wild of marine mammals that have spent significant time in captivity,[88] recognizing the need to prevent the spread of disease and release of unfit animals. Some animal protection advocates may propose MMPA amendments authorizing oversight of rehabilitation and release activities, requiring federal agency definition of rehabilitation/release protocols,[89] and establishing a scientific research permit for rehabilitation and release activities as well as for establishing rehabilitation/release facilities for long-captive marine mammals.[90] Such facilities might also engage in captive rotation programs, where animals are brought into captivity for predetermined amounts of time or are maintained in enclosures where they have periodic access to the open ocean. Proponents contend that the existence and operation of such facilities under strict guidelines would promote the welfare of captive and free-living marine mammals, including threatened and endangered species. Some public display interests and a few scientists, however, assert that rehabilitation and release does not work.[91] These critics cite research indicating that animals held in captivity for any length of time and those born in captivity are more likely to die upon release because they do not or are not able to make the necessary adjustments to life in the wild. They would oppose efforts that encourage the release of long-captive animals. Other opponents include those worried about the federal cost of financing such a program. A parallel concern relates to discouraging and preventing unregulated releases of captive marine mammals by the more proactive animal protection advocates.

Quality of captive environments

Under present MMPA regulations, captive marine mammals can be relocated anywhere that complies with APHIS regulations on captivity enclosure characteristics (e.g., bare concrete tanks are acceptable). Some scientists and animal protection interests assert that the captive environment of some U.S. marine parks is almost devoid of the features, richness, or dimensions of the natural world[92] of marine mammals — social animals that have evolved to exploit the complex and expansive natural marine environment.[93] Furthermore, they claim that our increased understanding of the complex social, psychological, and behavioral requirements of marine mammals reveals how lacking most captive environments are in providing sufficient space for animals to make normal postural and social adjustments or in allowing adequate freedom of movement. These critics would like the MMPA to be amended to require APHIS to define minimum acceptable levels of environmental and social stimuli for marine mammals. The physical and social environment of any animal regulated by the MMPA, it is argued, should conform to some standard for what is minimally acceptable and strive for enrichment to fulfill animals' needs. However, establishing standards to respond to the differing requirements of various species may be complex. For example, while some contend that overall size of the captive environment is much more important than its features for cetaceans, others believe that pinnipeds require more emphasis on geotopical elements in their artificial habitat rather than a large enclosure. In addition, it may be difficult or impossible to provide situations in captivity that permit the complex social systems, groupings, and bonding normal among marine mammals.

Programs promoting human interaction with captive dolphins

Various facilities holding captive dolphins promote interactive petting and feeding pools as well as programs for swimming with or wading with these animals. Animal protection advocates as well as some scientists and managers claim that these programs place both dolphins and humans at risk, and believe that APHIS regulation of such activities is inappropriately minimal. Early in 1999, APHIS suspended enforcement of all AWA regulations dealing with "swim-with-the-dolphin" programs to solicit further public comment on expanding regulations to encompass activities involving shallow water interactive programs with dolphins.[94] Some animal protection interests would like the MMPA and/or AWA to either prohibit all interactive programs, including petting and feeding pools which they claim have never been regulated, or require more stringent regulation of these programs by APHIS. These critics also suggest an inconsistency in policy and confusion of the public wherein swimming with and feeding of wild dolphins is prohibited to protect them from harassment while swimming with and feeding of captive dolphins, which could be less able to escape interaction, is promoted by marine parks. Those conducting interactive programs, however, argue that their activities are safe and well-managed, with adequate measures enforced to protect both dolphins and humans.

Insurance requirement

Since 1990, at least 21 North American marine parks are reported to have closed. Animal protection advocates suggest that measures need be taken to assure that the welfare of captive marine mammals is protected should research programs terminate or parks close. These interests may propose amending the MMPA to require that a minimum of $25,000 per marine mammal be placed in escrow or be covered by insurance as an additional permit requirement for each marine mammal transfer, import, and export. In addition, such a requirement might be imposed in permits covering each marine mammal born in captivity. Such financial resources would be used if the federal government were required to assume temporary responsibility for animals from closed parks or pay transfer expenses for moving animals to new facilities.

Prohibition of traveling exhibits

Animal protection advocates believe that circuses and traveling shows cannot maintain the highly specialized conditions necessary to ensure the health and well-being of marine mammals. They cite the recent experience with the Mexican-based Suarez Brothers Circus in Puerto Rico, where performing polar bears were confiscated by FWS. Dolphin traveling circuses exist and move throughout Latin America and the Caribbean, and could potentially enter U.S. territories or use marine mammals from U.S. facilities. Animal protection groups seek to amend the MMPA to prohibit these traveling exhibits.

Prohibition of wild captures for public display

The International Union for Conservation of Nature and Natural Resources' *Dolphins, Whales, and Porpoises: Conservation Action Plan for the World's Cetaceans, 2002-2010*[95] notes that the removal of live cetaceans from the wild for captive display is equivalent to incidentalor deliberatekilling, as the animals brought into captivity (or killed during capture) are no longer available to contribute to maintaining their populations. Concerned that, when unmanaged and undertaken without a rigorous program of research and monitoring, live capture can be fatally stressful to animals and pose a serious threat to cetacean populations, animal protection interests support an amendment to the MMPA to prohibit wild captures of marine mammals for public display.

Native Americans and Marine Mammals

Co-management with Native American tribes

Some federal managers believe that co-management agreements, when accompanied by dedicated funding, have dramatically improved communication among Native subsistence users, Alaska Native organizations, and FWS. However, Native Alaskan interests assert that NMFS has been slower to enter into cooperative agreements to implement co-management for marine mammals in Alaska (authorized under 16 U.S.C. §1388), and that federal appropriations to provide grants to Native organizations under this section have not been forthcoming.[96] Some Native American interests are likely to propose amending the MMPA to provide additional opportunities for Native Americans to participate in co-managing marine mammal populations, especially those that have subsistence value. Particular need is seen for coordinating federal and Alaska Native priorities in the Bering Sea region, due to ongoing concerns to better understand this marine ecosystem's apparent decline. Countering the view in support of additional co-management opportunities are some in the scientific and environmental communities who fear the potential for overhunting by Natives seeking economic gain, and who believe that current MMPA co-management provisions are more than adequate (if not excessively lenient). These critics believe co-management works well only when the federal government supports a multi-year national program to assess population abundance, habitat conditions, and ecological relationships to provide a sound basis for such co-management, as has been done since the 1970s for bowhead whales. Similar national programs have not been conducted on most other species. Some animal protection advocates are concerned that reporting of subsistence kill levels often lags by five years of more and is based on self-reporting, making it difficult to determine the impact of the subsistence on a particular stock until well after the fact. Animal protection interests also believe current cooperative agreements lack some transparency and provide little opportunity for public comment before the agreement is negotiated.

Reporting subsistence takes

Knowledge of the number of animals killed is necessary for managing any harvested resource. Nevertheless, many marine mammal stock assessment reports lack substantial information on subsistence takes. In 16 U.S.C. §1379(i), the MMPA states that "the Secretary may prescribe regulations requiring the marking, tagging, and reporting of animals taken pursuant to section 101(b)." FWS has promulgated regulations and instituted a marking, tagging, and reporting program (MTRP) for polar bears, walrus, and sea otters taken by Alaska Natives.[97] NMFS does not have a similar program, even though comparable information could be useful for managing species of special concern such as Steller sea lions and harbor seals.[98] Although NMFS has awarded contracts for the development of harvest estimates, their accuracy has been questioned by some scientists.[99] Some Alaska Native organizations conduct biosampling programs on marine mammals taken for subsistence through cooperative agreements developed under the authority of 16 U.S.C. §1388. Despite this, some in the Alaska Native and environmental communities continue to call for NMFS to develop an MTRP similar to that conducted by FWS, desiring more research on marine mammals taken for subsistence use.

The Alaska Native community generally accepts the FWS program, considering it to be well-run and to provide useful data. However, some managers and environmental interests believe the level of detail available on subsistence takes for many Alaska species could be improved. In particular, some animal protection interests, scientists, and managers do not consider the FWS program "well-run" and would like to see this program improved. Some scientists believe that a program for each species should include a well-designed harvest survey based on structured hunter samples from different communities that intensively exploit the targeted species, with data analysis by good statistical methods to adequately fulfill management needs. Other scientific and environmental interests suggest that the MMPA be amended to require reporting, marking, tagging, and sampling of all marine mammals taken by Alaska Natives for subsistence.[100] Others in the environmental and animal protection communities believe such reporting should be required for seal hunting and for anysubsistencetakes of marine mammals by Native Americans in the contiguous states (e.g., Washington, Oregon, and California). Some scientists, however, contend that tagging of subsistence kills may not be practical for species taken in large numbers, such as some seals, and that the sheer volume of individuals' subsistence activities may lead to under-reporting. In addition, some scientists and managers believe that better subsistence estimates need to be factored into the PBR process (see "Calculating Potential Biological Removal"), especially in situations where (1) subsistence harvest may account for the majority of the total number of animals removed and (2) subsistence harvest may approach or exceed the PBR level.[101]

Limitation on the sale of edible subsistence takes

In 16 U.S.C. §1371(b)(2), the MMPA states that "any edible portion of marine mammals may be sold in native villages and towns in Alaska or for native consumption." There are legitimate reasons why Alaska Natives purchase legally taken parts of marine mammals for their consumption.[102] The current interpretation of the MMPA language is that all Alaska locales, including the city of Anchorage, qualify as Native villages and towns. Certain markets in Anchorage sell large quantities of marine mammal meat and muktuk,[103] with a few Alaska Natives allegedly hunting primarily to supply this commercial market.[104]

Scientists, animal protection advocates,and environmentalists suggest amending the MMPA to limit or restrict the sale of edible parts from marine mammals taken for subsistence, such as prohibiting commercial sales in cities or in communities where Native residents are in the minority. Others suggest amending the MMPA to prohibit the commercial sale of marine mammal products from any stock that is declining in abundance. Alternatively, NMFS and/or FWS already have the authority to make administrative determinations that species are depleted under the MMPA or are threatened/endangered under the ESA, allowing them to take regulatory action to limit subsistence take without legislation.[105]

Alaskan Natives, however, believe that the Native community itself should take the initiative to deal with these problems, using existing models that have proven workable in similar Alaska Native situations. They suggest approaches similar to those used in the allocation of strikes[106] among various whaling crews in the North Slope Borough or the Sitka Tribe's management of sea otter take in traditional territory.[107] Others are concerned about the potential cultural costs of limiting access to subsistence foods for individuals living in urban areas and the possibility that these costs could outweigh the benefits to marine mammal stocks.

Definition of subsistence whaling

With the support of the U.S. government, the Makah Tribe of Washington State petitioned the International Whaling Commission (IWC) in 1996 for an allocation to harvest eastern Pacific gray whales, to exercise whaling rights as part of their cultural heritage negotiated in the 1855 Treaty of Neah Bay between the Makah and the United States. In October 1997, a bilateral agreement between Russia and the United States on aboriginal quota sharing resulted in the Makah gaining access to IWC aboriginal quota sufficient to kill an average of four gray whales from the North Pacific stock annually from 1998 through 2002.[108] Disagreement continues, both domestically and internationally, concerning the appropriateness and legitimacy of the action taken on this issue.[109]

While bowhead whaling by Native villagers along Alaska's Beaufort and Chukchi Sea coasts is seen as truly for the subsistence, animal protection advocates are concerned that the Makah seek to kill whales without demonstrable proof of nutritional need, but with an eye to the possibility of commercial trade in whale products. To animal protection interests, this has the potential for reversing the whale's recovery and for inviting a return to whaling by all northern cultures which claim whaling as part of their cultural tradition. In fact, after the Makah situation, Native peoples in Canada demanded their "cultural right" to return to whaling. Although Norwegians, Icelandics, Faroese,Irish, Japanese, Russian, and others assert cultural traditions in whaling, their situations and that of Canadian aboriginal groups differ from the Makah in that no "right towhale" has been acknowledged by treaty.[110] Animal protection and some scientific interests suggest amending the MMPA to make a clear distinction between non-subsistence and subsistence whaling and to establish more stringent criteria for non-subsistence whaling, allowing only minimal token quotas/takes of those stocks determined to be fully recovered. Others suggest the MMPA be amended to require that the United States take no action that might "diminish the effectiveness" of the IWC, similar to languagein thePellyAmendment to the Fishermen's Protective Act (22 U.S.C. §1978) that is applicable to foreign nations with whom the United States trades. However, it is uncertain whether Congress has the authority to take any action that might alter or limit the terms of the 1855 Treaty of Neah Bay.

Definition of subsistence

Several parties suggest that policy relating to "subsistence" is confused and needs clarification, requiring attention to both ethics/tradition and biology/ecology for resolution. Some of the confusion was created when the MMPA waived the moratorium on taking of marine mammals by Alaska Natives, placing federal and Alaskan law and regulations in conflict.[111] This confusion was exacerbated by the interaction of western technologies and economies on traditional beliefs and practices. For example, reported annual walrus kills for the St. Lawrence Island communities of Gambell (1,300 animals) and Savoonga (700 animals), composed mostly of females, appears excessive and questionable as "subsistence" to some managers, scientists, and animal protection groups. FWS regulations on the use of meat, skin, etc., are minimal and result in significant waste in a harvest that focuses on obtaining ivory. Some scientists and managers suggest that the MMPA be amended to base subsistence policy more firmly within the context of a species' biological and ecological requirements, with social/cultural values taken into secondary account within that framework.

Cultural exchange

While the 1994 MMPA amendments appeared to have improved cultural exchange among Inuit peoples as far as imports of marine mammal products by Alaskan Natives are concerned, problems remain with the export of marine mammal products by Alaska Natives for these purposes. In addition, problems arose in July 1999 when handicraft whalebone and sealskin marionettes used in portraying traditional Inuit legends were intercepted and seized by the U.S.

Customs Service as violating the MMPA. The marionettes had been shipped by Canadian Inuit to a U.S. craftsperson for finishing-detail adjustments. Native and some scientific interests suggest that the MMPA might be amended to be less restrictive of cultural exchanges involving marine mammal products.

Permits and Authorizations

Polar bear sport hunting in alaska

After the 1994 amendment of the MMPA to permit the import of polar bear trophies from Canada,[112] the sport hunting community may seek further amendment to allow polar bear sport hunting in Alaska under a strict, conservative quota. Proponents of such an amendment suggest that this action might promote better polar bear management and could result in additional funding for polar bear research and management. The animal protection community almost certainly would oppose such a proposal, and some may even seek repeal of the 1994 amendments allowing the import of polar bear trophies from Canada. Animal protection advocates substantively disagree with the theory that sport hunting promotes sound or sustainable management and that quotas in Canada's hunts are strict or conservative.[113] In early 2007, FWS proposed that polar bears be listed as threatened species under the Endangered Species Act.[114]

Large incidental takes

MMPA provisions (16 U.S.C. §1371(a)(5)(A)) authorize federal managers to issue permits for U.S. citizens to incidentally take small numbers of marine mammals.[115] However, the MMPA lacks a comparable program to deal with large incidental takes, other than those by the commercial fishing industry (for more information, see "Commercial Fishing Interactions with Marine Mammals"). Related to this, the regulatory burden for protecting marine mammals appears to fall inequitably on different industries. For example, while small incidentaltakepermits are regularly required by NMFS for offshoreoil and gas exploration and development activities, NMFS does not regulate large commercial vessel traffic under the small incidental take program,[116] despite concerns that serious injury and mortality of cetaceans due to vessel strikes may be significant. Other activities that may "take" large numbers of marine mammals by harassment include whale-watching vessels, high-speed ferries, recreational jet skis, and other sources of anthropogenic noise. By statute, small take permits may be issued for periods of as long as five years under regulations, or one year under incidental harassment permits, with congressional report language indicating an intent that such permits be renewable. NMFS claims that most permits limit taking to small numbers of animals[117] by harassment because mitigation measures imposed by NMFS on the activity prevent serious injury or mortality to marine mammals. If it were proposed that the MMPA be amended to address this issue, individuals who might be required to comply with these modified permitting procedures (e.g., jet skis, whale-watching vessels, ocean transport vessels) would likely oppose such a change if the new requirements were viewed as imposing additional or burdensome restrictions on their activities. Some environmentalists and animal protection advocates recognize that the permitting process is a relatively inefficient way to mitigate impacts from vessel traffic and suggest that a separate management scheme, protective of marine mammals, would be more appropriate to address both vessel-strike and anthropogenic noise concerns.

Noise and its effects

Noise as a category of potential harm to marine mammals is unique in that sound propagates both horizontally (near/at the surface) and vertically (down to substantial depths). Anthropogenic acoustics (e.g., ship traffic, military active sonar, seismic exploration, explosives trials, acoustic harassment devices used by fishermen) permeate the water column and have the potential to affect numerous unseen marine mammals, fish, diving birds, and other marine life. Although it is difficult to measure the potential that noise has to harm or harass unseen animals, the U.S. Navy, the Minerals Management Service, and other agencies have invested considerable time and funds on research to develop monitoring capabilities and to document and quantify the impact from specific noise sources on certain species under known conditions.[118] However, significant information remains lacking on sound impacts on cetaceans, on behavioral and physiological reactions of marine mammals, and on which species are exposed at what depths and distances from sound sources. For this reason, the effect of noise on marine mammals is subject to much speculation, presumption, and misinformation. FWS and NMFS have reacted to issue-or site-specific concerns, generally through the permit process, but they have not issued any guidance or regulations concerning anthropogenic noise, nor have they implemented any systematic monitoring or enforcement programs.

Some scientists,[119] believing that the benefits of acoustic research may outweigh any potential effect on marine mammals, may propose amending the

MMPA to simplify procedures for federal authorization of incidental taking from acoustic noise. As one approach, these scientists suggest that the MMPA might be amended to authorize the regulation of impacts collectively as broad categories or classes of sound-producing activity rather than separate individual actions.[120] Other proposals might include revising the definition of level B harassment (16 U.S.C. §1362(18)(A)(ii)) to be applicable to actions that can reasonably be expected to constitute a significant threat only to marinemammal stocks ratherthanalso to small numbers of individual animals. Reasons offered by some in the scientific community for change include (1) some of the most prevalent anthropogenic noisemakers, including personal watercraft (e.g., jet skis), large high-speed oceangoing ships, and whale-watching vessels, are unregulated;[121] (2) a disproportionate "harassment" burden is placed on scientists using acoustics for research (i.e., direct research into the potential effects of sound on marine life is subject to a higher regulation and compliance burden than any other human-made ocean acoustic activity); and (3) human-made sound in almost all cases is neither as loud nor as constant as naturally-occurring ocean activity (e.g., subsea earthquakes, rain on the sea surface, volcanic eruptions, and whale calls themselves).

The effects on marine mammals by active sonar development and deployment by the military has been of intense concern. Coincident with low-frequency active (LFA) sonar tests conducted by a NATO research vessel in the vicinity, a mass stranding and death of 12 Cuvieri's beaked whales was observed in May 1996 in the eastern Mediterranean Sea (Ionia Sea).[122] The mass stranding of at least 15 whales of four species (at least 7 of these animals died) in the Bahama Islands on March 15, 2000, occurred coincidental to U.S. Navy transit and activities in the area.[123] Additional strandings of beaked whales have been observed in conjunction with mid frequency active sonar exercises in Madeira (2000) and the Canary Islands (2002). A September 2002 beaked whale stranding in the Gulf of California occurred concurrently when a vessel operated by Columbia University's Lamont-Doherty Earth Observatory pulsed the ocean with high-powered sound waves to map the lithosphere beneath the ocean floor.[124] More than five years of regulatory attention to deployment of low frequency active sonar by the U.S. Navy, with accompanying legal challenges, culminated in publication of a final rule in July 2002,[125] with letters of authorization required for subsequent deployment.[126] In addition and in recognition of concerns raised in federal court[127] over use of the LFA system and to further its commitment to responsible stewardship of the marine environment, the Navy is preparing a supplemental environmental impact statement on this technology.[128]

Some animal protection advocates, environmentalists, and scientists characterize many sources of anthropogenic noise in the ocean as increasingly persistent and regular. These critics point to a growing body of evidence, particularly the mass mortalities of beaked whales associated with military active sonar use, as indicative that current mitigation practices are insufficiently protective of marine mammals. Believing that too little is known about the long-term effects of noise on marine mammals,[129] these critics believe a precautionary approach is necessary and oppose any action that could be interpreted as liberalizing the regulation of anthropogenic sources. In addition, these critics are especially concerned with low-frequency sound that is produced at very high pressure levels and is designed to travel thousands of miles through the ocean, as opposed to other anthropogenic noise that dissipates relatively quickly in the ocean. Furthermore, these critics suggest that, rather than exempting acoustic scientists from regulation and permitting additional sources of ocean noise, other sources of non-research-related noise should be more aggressively regulated to reduce this harassment. A variety of constituencies[130] might support a proposal to authorize and fund a major research effort directed at increasing understanding of the potential effects of anthropogenic noise sources on marine mammals.

Research permits for NMFS and FWS scientists

The MMPA (16 U.S.C. §1374(c)(3)(A)) provides a lengthy process for issuing scientific research permits. NMFS and FWS are funded by Congress to study marine mammals as necessary to provide a sound basis for their conservation and management. Some federal scientists would like to see the MMPA amended to facilitate federal research on marine mammals by eliminating the cumbersome process of obtaining scientific research permits. These federal researchers question the necessity of requiring federal agency personnel to request permits from another part of their own agency before they can do their work. These critics suggest that the MMPA be amended to provide scientists within the federal management agencies with a blanket authorization for research. Others suggest that relief from the lengthy permitting process be extended to all those involved in conducting federally funded research. This could include an exemption from permits or an expedited permit review procedureoffering a simpler issuance or renewal of permits for studiesunchallenged by public comment. It might also be applicable to state wildlife agencies when their scientists work in direct cooperation with one of the federal agencies. Some nonfederal scientists, animal protection advocates, and environmentalists argue that regular reporting as well as outside peer and/or public review are especially necessary for government scientists who could be influenced by political considerations. They also would object to preferential treatment of federal researchers as discriminatory, arguing that federal research should be required to meet the same standards, requirements, and scrutiny as non-federal research.[131]

Scientific research permits

Several issues concern the administration of scientific research permits by federal management agencies. Some scientists criticize FWS and NMFS permit offices for delays in processing requests for scientific research permits, even though the MMPA mandates a 30-day deadline for agency action.[132] While some permits are processed quickly, others may take many months longer, they charge, with no explanation or obvious differences between them. Critics report that the delay between submission of a permit application and its publication in the *Federal Register* for public comment can be six weeks or more.[133] To assist the agencies in expediting the permit review process, they suggest that the MMPA be amended to authorize committees of scientists that would review scientific research permit applications in the same fashion that committees review proposed research on human and animal subjects.[134] Such committees might also be helpful in addressingconcerns about alleged misuse of scientific research permits by whale-watching operators, dolphin encounter tour brokers, and others wherein "paying volunteers" are recruited to help conduct "research" of questionable value. Although this latter issue could be addressed administratively, some scientists believe congressional direction might be helpful or even necessary if administrative action is not forthcoming. Scientists are also concerned with permit restrictions that they interpret as constraining their ability to conduct manipulative and invasive research on marine mammals, albeit with adequate safeguards.

Some scientists suggest that the entire scientific research permit process needs to be streamlined, especially what are seen as (1) restrictive, burdensome, and unreasonable procedural requirements (i.e., level of specificity and amount of paperwork) related to justify level B harassment (see also the discussion of "Harassment") for *bona fide* research; and (2) unduly tedious and specific requirements of the annual reporting process. Scientists feel they are subjected to a much more stringent regulatory regime (e.g., see also the discussion of "Noise and Its Effects") than is imposed on activities that appear to be potentially more harmful to marine mammals (e.g., commercial fishing).

On the other hand, animal protection advocates assert that permit processes and requirements are not too restrictive when it comes to invasive research, and both they and environmentalists argue that there is little justification for treating the research community as privileged. While certain amendments might streamline the permitting process for certain research with a low harassment potential or to establish streamlined programmatic permitting for certain kinds of research, the environmental and animal protection communities both disagree that research should be seen as having less of an impact on the marine environmental and marine mammals as a general matter when compared to fishing or other human activities.

State approval of federal MMPA permits

Under the authority of the federal Coastal Zone Management Act, three states (Hawaii, Washington, and Alabama)[135] include in their state coastalplans the requirement that the state approve federal permits granted under the authority of the MMPA. Some scientists are concerned that state review of federal marine mammal permits is duplicative and burdensome for marine mammal researchers and circumvents the procedures in the MMPA (16 U.S.C. §1379) for granting state management authority over marine mammals.[136] These critics suggest that Congress may wish to review whether this action improves protection for marine mammals.

Program Management and Administration

Definition of "Take."

Some scientists suggest it might be worthwhile to re-evaluate the MMPA definition of *take* in their belief that the current definition may be overly broad and encompassing, as well as unenforceable in many situations. These critics suggest that the MMPA be amended to incorporate a new definition of *take* that establishes an enforceable, biologically significant standard for interactions with individual marine mammals (for ESA-listed and depleted species) and for marine mammal populations (for all other species). With such a standard, they argue, management will focus specifically on interactions which are likely to have adverse biological significance for these animals. However, animal protection advocates and environmental groups might be anticipated to oppose any effort to redefine take that might be perceived as reducing the scope of activities prohibited or regulated under the MMPA.

Trade in marine mammal parts and products

The U.S. government has experienced pressure fromthe World Trade Organization (WTO) regarding the trade barriers inherent in many U.S. environmental statutes. Importing marine mammals and their products into the United States is prohibited by 16 U.S.C. §1371(a), except under special permits for scientific research, public display, photography for education or commercial purposes, or enhancing the survival or recovery of a species or stock. Permits also may be granted to import polar bear parts, other than internal organs, taken in legal Canadian sport hunts. In addition, the ESA and the Convention on International Trade in Endangered Species of Wild Fauna and Flora (CITES, the international agreement implemented through the ESA) impose additional restrictions on trade of certain listed marine mammals. Given the desire of several nations to commercially trade in marine mammal products (particularly whalemeat and pinniped products), some suggest that Congress could act to possibly forestall a WTO challenge to U.S. prohibition of such trade by amending the MMPA to allow limited trade[137] or in some other limited manner to make the MMPA more compatible with WTO rules.[138] Such a proposed change would likely be vigorously opposed by some in the environmental, scientific, and animal protection communities who fear that opening U.S. markets could promote increased kills in nations less protective of marine mammals.[139] They further assert that, if such a proposal were enacted, U.S. policy would be inconsistent, prohibiting domestic commercial exploitation of marine mammals while encouraging or allowing foreign commerce in these same protected animals' products in the United States. They also argue that the availability of foreign marine mammal products on the U.S. market could encourage the illegal harvest of domestic marine mammals for these same markets. An alternative, although likely more difficult, approach seeks to broaden WTO rules such that the MMPA could be found compatible.

Management of robust stocks

Populations of California sea lions and Pacific harbor seals have been increasing along the Washington, Oregon, and California coasts, leading to more frequent interactions between these animals and fishermen and the general public (particularly the marina/boating public). On February 10, 1999, in response to the requirements of 16 U.S.C. §1389(f), NMFS delivered an 18-page report to Congress and released a supporting 84-page scientific document on management conflicts related to rapidly increasingpopulations of West Coast harbor seals and California sea lions.[140] How to manage these stocks is expected to be an issue during MMPA reauthorization. The issue is seen by some as whether an increasing human population on the West Coast can co-exist with a truly robust pinniped population or whether these pinnipeds will be adversely affected by coastal development and marine resource exploitation or will themselves have an adverse effect on coastal resources.[141]

While some local residents and fishermen fear that these pinniped stocks may be "over-populating," scientists, environmentalists, and animal protection advocates counter that populations may be merely returning to their historic carrying capacities after over-exploitation diminished their abundance earlier in the 20th Century. Fishing industry or local government officials may propose that the MMPA be amended to permit selective culling or additional lethal nuisance animal control. NMFS has recommended that Congress consider amending the MMPA to create a new framework thatwould allow state and federal resource managers to immediately address site-specific conflicts involving California sea lions and Pacific harbor seals.[142] In this chapter, NMFS suggests that a streamlined approach provide procedures for lethal removal of these species where they are harming severely depleted salmonids (including some populations listed as threatened or endangered under the ESA), where they are harming salmonid populations identified as being of special concern by states, and where they are in conflict with human activities. While the MMPA in 16 U.S.C. §1389 already provides for the lethal removal of pinnipeds to protect human safety and fish stocks, federal and state managers view the process for implementing the existing provisions as lengthy and overly cumbersome. Environmental, animal protection, and scientific critics, however, condemn the idea of culls and lethal nuisance animal control as excessive. They believe that such an approach deflects resources from addressing other expensive and contentious human activities that contribute to fish stock declines (e.g., habitat degradation, siltation, water diversions, fish passage at dams, overfishing)[143] and that non-lethal deterrents have not been adequately explored. Furthermore, they express concern that authorizing the killing of marine mammals interacting with wild fish stocks appears counter to the MMPA's mandate to manage on an ecosystem basis.[144] In addition, these critics are adamant that nuisance animal control not be authorized for human activities (e.g., aquaculture) that can and should be sited so as to avoid areas of potential conflict with marine mammals.

Fostering international cooperation

Although the MMPA established an international program (16 U.S.C. §1378), little framework exists to foster international cooperation between the

United States and foreign countries on marine mammal issues. Under the MMPA, international cooperation — funding, exchange programs, and cooperative research — has been limited largely to the industry-centered dolphin/tuna issue.[145] MMPA funds for such activities are at least an order of magnitude less than the millions of dollars in federal U.S. endangered species funds that are used to foster international cooperation to protect elephants, tigers, rhinoceros, great apes, and other species.[146] Proponents of increased international cooperation argue that no similar program for marine mammals is provided in the MMPA or elsewhere in U.S. law. For example, although the U.S.-managed North Atlantic right whale is endangered and its population is not rebounding, the southern right whale population is flourishing. A Brazilian right whale project focuses on reducing human/whale interactions where ship strikes have been a major cause of death. Cooperative activities that might be promoted include sharing whale monitoring and collision avoidance procedures as well as whale reproduction, health, and population information with Latin American authorities. Marine mammal scientists suggest that Congress may want to consider the benefits of encouraging international cooperative relationships on marine mammals by U.S. agencies. Most marine mammal constituencies appear supportive of efforts to encourage more international cooperation, as long as such action does not promote invasive research or commercial ventures.

Some also suggest there may be a critical need for expanding international cooperative programs for Arctic species because of the virtual total demise of Russian research and management programs, and the increasing pressure by protein-impoverished Native peoples to take these species for subsistence purposes. Many of these species are "shared" because their ranges include the waters of both the United States and the Russian Federation. A decline of Russian management effort has hampered population assessment programs for these shared species.

Critics, however, warn that, if Congress acts in this area, specific language might need to be incorporated to prevent potential abuse (i.e., expenditure of funds intended to recover and protect U.S. marine mammal stocks on questionable studies of exotic marine mammals in interesting places) and to require appropriate guidance and accountability to ensure that international efforts are reciprocal and relevant.

Harassment

The 1994 MMPA amendments revised the definition of *harassment* to distinguish between two levels of interaction — those with the potential to injure (level A harassment) and those with the potential to disturb (level B harassment).[147] Some federal managers have found the new definition of level B harassment to be particularly difficult to enforce,[148] and potentially harmful human interaction with marine mammals continues.[149] Other critics suggest whale-watching vessels are insufficiently monitored for compliance with MMPA regulations.[150] Animal protection advocates suggest that the MMPA should be amended to require specific and more strictly enforced regulations concerning swimmer,[151] kayaker, and boater harassment of dolphins and whales, including provisions to significantly increase the possible fines against commercial operators who introduce large groups of swimmers into protected bays where dolphins rest. Others suggest authorizing more funding specifically targeted to better educate private watercraft operators concerning MMPA regulations and to increase MMPA enforcement efforts,[152] including additional observers aboard whale-watching vessels to assess compliance. Some scientists, on the other hand, would like to see the definition of level B harassment revised to where it would be applicable only to situations where actions would reasonably be expected to constitute a significant threat to an entire marine mammal stock, rather than to just a few individual animals.

Changes to the harassment definition applicable to military readiness operations and to scientific research activities conducted by or on behalf of the federal government were enacted in §319(a) of P.L. 108-136. The new language defines harassment as any action that "injures" or "has the significant potential to injure" marine mammals, rather than any action that has the "potential to injure." Environmental and animal protection organizations generally oppose the modified definition of harassment, arguing that it raises the burden of proof that a military readiness activity would affect a marine mammal, making it more difficult to protect them.[153] These interests believe that such changes are premised on an unrealistically high assessment of our ability to differentiate between biologically significant and insignificant responses. By doing so, they believe the modified definition effectively reverses the precautionary burden of proof that has been the hallmark of the MMPA since its inception. Supporters of the modified definition believe that it ensures that activities are restricted only when scientific evidence demonstrates that such protection is necessary. These changes remain highly controversial and could be revisited during MMPA reauthorization.

Management consistency between FWS and NMFS

The division of responsibility for various marine mammal species between NMFS and FWS is provided for in 16 U.S.C. §1362(12) within the definition of "Secretary." When the MMPA was enacted in 1972, this division of species was seen as artificial and temporary by many in Congress and the Administration, awaiting the creation of a contemplated "Department of Environment and Natural Resources." In addition, the differing management approaches taken by NMFS and FWS have often confused the commercial fishing industry and Alaska Natives.[154] Some Native American and scientific interests suggest that it may be time for Congress to revisit this division of management responsibility and consider amending the MMPA to promote greater consistency in marine mammal management. Several approaches are suggested, including the current movement toward an ecosystem approach to managing living resources and minimizing possible conflicts of interest where marine mammals and fisheries interact, that may have a bearing on which agency should manage which species or groups thereof. Others suggest that NMFS and FWS might be directed to develop joint regulations for all their marine mammal programs to achieve greater consistency in management policy.[155]

Directed research program

Although the MMPA emphasizes research, it does not create a national integrated marine mammal research program. Emphasizing this need, a recommendation in the Secretary of Commerce's February 1999 report to Congress included a list of information needs, with no suggestion as to how or by whom this research was to be pursued.[156] Specific information needs identified for this relatively narrowly focused issue include (1) site-specific investigations on the impacts of pinniped predation on salmonid populations; (2) state-by-state and river-by-river investigations of salmonid populations vulnerable to pinniped predation; (3) studies of comparative skeletal anatomies of different salmonid species so that prey may be identified in food habit studies using pinniped scat and gastrointestinal tract analyses; (4) site-specific seasonal abundance and distribution of pinnipeds north of Point Conception, California; (5) assessment and evaluation of potential impacts of pinnipeds on specific fisheries and fishing areas; (6) socioeconomic studies on impacts of pinnipeds on various commercial and recreational fisheries; (7) ecosystem research where the impacts of pinniped predation on non-salmonid resources can be addressed beginning with smaller systems such as Puget Sound, Washington; and (8) collection of unbiased samples for food habit studies. Some have suggested that Congress might wish to consider whether these information needs should become the focus of an MMPA amendment creating a national integrated research program, possibly under the direction of the independent Marine Mammal Commission, with specific funding authorized.[157]

Federal agency roles

Some scientists suggest that conflicting federal agencyinterests may hamper marine mammal protection and recovery. One example of an interagency issue where conflicting agency authority may be problematic relates to understanding and addressing the potential for endocrine disruption in marine mammals.[158] Some critics suggest that the MMPA be amended to direct an external panel (e.g., the Marine Mammal Commission or the National Academy of Sciences) to carefully review the programs and procedures of federal management agencies for potential conflicting interests among their management, regulation, permit administration, scientific research, and funding roles with respect to marine mammals, and recommend actions that should be taken to address any problems identified.

Agency delays in compliance with MMPA deadlines

Various constituencies were frustrated over federal agency delays in implementing provisions of the 1994 MMPA amendments (see "1994 MMPA Reauthorization" for more detail). This led to critics within the conservation and animal protection communities as well as the fishing industry to seek additional means to force NMFS and FWS to comply with MMPA deadlines.[159] These agencies contend, in reply, that the problem can be traced to limited funds provided by Congress to finance these activities.[160] For more information on funding concerns, see the following section "Appropriation of Agency Funding." Others suggest that the pattern of repeated failure to complete assigned tasks on time should first be addressed through an Office of Management and Budget or similar study on overall agency administration.

Appropriation of agency funding

An issue that NMFS, FWS, and the MMC might raise during reauthorization discussions is that, in an era of stable or slightly declining federal appropriations, federal agency responsibilities and duties have expanded much faster than their budgets. For example, requirements for stock assessments of all marine mammal populations, observer monitoring aboard the U.S. commercial fishing fleet, and administration of an expanding MMPA permit program place significant demands on federal agency resources. Specifically, implementing the provisions of the MMPA is a significant undertaking, requiring the coordination among headquarters and regional offices of NMFS and FWS as well as with the MMC. NMFS and FWS may claim that they are underfunded and that delays in implementing the 1994 MMPA amendments were directly tied to budgetary constraints. In addition, some scientists, environmental interests, and animal protection advocates assert that Congress' will to implement the MMPA through the appropriations process has not kept pace with the desire expressed in the authorization process, providing insufficient funding for federal management programs and thereby exposing federal agencies to harsh criticism when they fail to meet public expectations. Some suggest that Congress might consider amending the MMPA to increase the authorization of appropriations for marine mammal programs of NMFS, FWS, and the MMC, increasing the actual funds appropriated under these authorizations, or including specific instructions about how funds are to be used.

CONGRESSIONAL OUTLOOK

Congress has enacted measures to protect marine mammals, including specifically the MMPA. While the history of the MMPA's implementation includes numerous court challenges to agency interpretation of congressional intent, Congress generally has been understanding of the difficulties in providing protection for this specific group of living resources. The issues discussed in this chapter set before Congress a varied array of concerns. It is not yet clear which will gain prominence in any comprehensive reauthorization debate. Recent public sentiment, always a strong factor in marine mammal issues, has focused on concerns about noise in the marine environment, enhanced protection for whales, the appropriateness of Makah whaling, and humane care for captive animals. Specific interests of Native Alaskans, fishermen, sport hunters, and animal protection groups may call attention to additional issues. Congressional oversight during the reauthorization process is likely to identify additional issues. The requirements of the MMPA itself did not expire when the authorization of appropriations expired at the end of FY1999. With the delay in enacting a comprehensive reauthorization, Congress has separately considered provisions to amend the MMPA on a number of selected issues.

During the 106th Congress, the House Resources Subcommittee on Fisheries Conservation, Wildlife, and Oceans held a general oversight hearing on the Marine Mammal Protection Act on June 29, 1999. Testimony presented by officials from NMFS, FWS, APHIS, and the MMC described progress in implementing the 1994 MMPA amendments and outlined possible areas for Committee attention during reauthorization. The same subcommittee held a second oversight hearing on April 6, 2000, specifically on the implementation of the 1994 amendments related to the *take reduction* process, cooperative agreements with Alaska Native organizations, and co-management of subsistence use of marine mammals by Alaska Native communities. No other action was taken on MMPA reauthorization in the 106th Congress.

In the 107th Congress, the House Resources Subcommittee on Fisheries Conservation, Wildlife, and Oceans held a general oversight hearing on October 11, 2001, on reauthorizing the Marine Mammal Protection Act.[161] H.R. 4781 was the only reauthorization bill that was introduced; the House Resources Subcommittee on Fisheries Conservation, Wildlife, and Oceans held a hearing on this bill on June 13, 2002,[162] and marked up this measure on July 25, 2002. No further action was taken.

In the 108th Congress, H.R. 2693 and H.R. 3316 would have amended and reauthorized the MMPA through FY2008. The House Resources Subcommittee on Fisheries Conservation, Wildlife, and Oceans held a hearing on H.R. 2693 on July 24, 2003; on April 20, 2004, the House Committee on Resources reported (amended) this bill (H.Rept. 108-464). On July 16, 2003, the Senate Commerce Subcommittee on Oceans, Fisheries, and Coast Guard held a hearing on MMPA reauthorization issues. On August 19, 2003, the House Resources Subcommittee on Fisheries Conservation, Wildlife, and Oceans held an oversight field hearing in San Diego, California, on the increasing frequency of interactions betweenmarinemammals and humans. H.R. 5104 would have amended the MMPA and authorized appropriations for the John H. Prescott Marine Mammal Rescue Assistance Grant Program through FY2009; this measure was reported by the House Committee on Resources on November 19, 2004 (H.Rept. 108-787).

In the 109th Congress, H.R. 2130 and H.R. 4075 would have extensively amended the MMPA and authorized appropriations for several programs; the House Committee on Resources reported H.R. 2130 (amended) on July 21, 2005 (H.Rept. 109-180). The House passed H.R. 4075 (amended) on July 17, 2006. Title IV of S. 1224 would have amended the MMPA to encourage development of fishing gear less likely to take marine mammals, expanded fisheries required to participate in the MMPA incidental take program to include recreational fisheries, and authorized appropriations for stock assessments and observer programs; in addition, Title III (Subtitle C) directed negotiation of international agreements to better protect cetaceans from commercial fishing gear and authorized a grant program to develop less harmful fishing gear. Section 206 of H.R. 2939 would have transferred management of all marine mammals to NOAA. H.R. 3839 would have amended the MMPA to repeal the long-term goal for reducing to zero the incidental mortality and serious injury of marine mammals in commercial fishing operations, and to modify the goal of take reduction plans for reducing such takings. H.R. 6241 would have amended the MMPA to authorize taking of California sea lions to reduce their predation on endangered Columbia River salmon.

For updated information on legislative activities in the 110th Congress concerning marine mammals, see CRS Report RL33813, *Fishery, Aquaculture, and Marine Mammal Legislation in the 110th Congress*, by Eugene H. Buck.

Oceans Commissions Reports

Two ocean commissions recently released reports relating to marine mammals. The Pew Oceans Commission report[163] was released June 4, 2003, and the U.S. Commission on Ocean Policy's preliminary report[164] was issued on April 20, 2004. Marine mammal issues are only one aspectof the comprehensive ocean policy issues discussed in these reports; the larger context includes governance, education, coastal development, human health, environmental quality, energy resources, and ocean science, among others. For background on the reports and the larger context of these issues, see CRS Report RL33603, *Ocean Commissions: Ocean Policy Review and Outlook*, by Harold F. Upton, John R. Justus, and Eugene H. Buck. **Table 1** summarizes these two reports' recommendations relating to marine mammals. CRS takes no position with respect to either report's recommendations.

Table 1. Ocean Commissions Recommendations Relating toMarine Mammals

Issue	U.S. Commission on Ocean Policy	Pew Oceans Commission
Marine Mammal Commission	Recommendation 20 — 1. Congress should amend the Marine Mammal Protection Act to require the Marine Mammal Commission, while remaining independent, to coordinate with all relevant federal agencies	No similar recommendation.

Table 1. (Continued)

Issue	U.S. Commission on Ocean Policy	Pew Oceans Commission
	through the National Ocean Council (NOC). The NOC should consider whether there is a need for similar oversight bodies for other marine animals whose populations are at risk.	
Agency jurisdiction	Recommendation 20 — 2. Congress should amend the Marine Mammal Protection Act to place the protection of all marine mammals within the jurisdiction of the National Oceanic and Atmospheric Administration.	Congress should establish a National Oceanic and Atmospheric Agency as an independent agency outside the Department of Commerce. This agency should include the marine mammal programs of the Department of the Interior to place all ocean wildlife under the jurisdiction of the oceans agency.
Harassment definition	Recommendation 20 — 5. Congress should amend the Marine Mammal Protection Act to revise the definition of harassment to cover only activities that meaningfully disrupt behaviors that are significant to the survival and reproduction of marine mammals.	No similar recommendation.
Research on effects of human activities	Recommendation 20 — 7. The National Oceanic and Atmospheric Administration and the U.S. Department of the Interior should promote an expanded research, technology, and engineering program, coordinated through the National Ocean Council, to examine and mitigate the effects of human activities on marine mammals and endangered species.	No similar recommendation.
Specifically, acoustics and noise	Recommendation 20 — 8. Congress should increase support for research into ocean acoustics and the potential impacts of noise on marine mammals. This funding should be distributed across several agencies, including the National Science Foundation, U.S. Geological Survey, and Minerals Management Service, to decrease the reliance on U.S. Navy research in this area. The research programs should be well coordinated across the government and examine a range of issues relating to noise generated by scientific, commercial, and operational activities.	No similar recommendation.
and toxics	No similar recommendation.	Sufficient resources should be devoted to studying the effects of toxic substances in the marine
		environment. Needed research

	·	includes the effects of polychlorinated biphenyls (PCBs) and other toxic substances on marine mammals — particularly in the polar regions.
Conservation decision-making	No similar recommendation.	Core conservation decisions should be made by NMFS, or a revamped fishery service within a new independent oceans agency. These decisions should originate at the regional offices with oversight by the national headquarters office. At a minimum, these decisions include setting specific protected species requirements (threatened and endangered marine mammals, sea turtles, seabirds, and fish).
Regulation of sound	No similar recommendation.	Activities that generate significant amounts of potentially harmful sound should be regulated consistent with the requirements of federal law, including the Marine Mammal Protection Act.
Permits:		
Streamline an interagency process	Recommendation 20 — 6. The National Marine Fisheries Service and the U.S. Fish and Wildlife Service should implement programmatic permitting for activities that affect marine mammals, wherever possible. More resource intensive case-by-case permitting should be reserved for unique activities or where circumstance indicate a greater likelihood of harm to marine mammals. The National Ocean Council should create an interagency team to recommend activities appropriate for programmatic permitting, those that are inappropriate, & those that are potentially appropriate pending additional scientific information. Enforcement efforts should also be strengthened and the adequacy of penalties reviewed.	No similar recommendation.
Clarify what activities require permits	Recommendation 20 — 4. Congress should amend the Marine Mammal Protection Act to require the National Oceanic and Atmospheric Administration to more clearly specify	No similar recommendation.
	categories of activities that are allowed without a permit, those that require a permit, and those that are prohibited.	

Table 1. (Continued)

Issue	U.S. Commission on Ocean Policy	Pew Oceans Commission
Attention to bycatch reduction	Recommendation 19 — 25. The National Oceanic and Atmospheric Administration, working with the U.S. Fish and Wildlife Service and the U.S. Department of State, should design a National Plan of Action for the United States that implements, and is consistent with, the International Plans of Action adopted by the United Nations Food and Agriculture Organization and its 1995 Code of Conduct for Responsible Fisheries. This National Plan should stress the importance of reducing bycatch of endangered species and marine mammals.	No similar recommendation.

End Notes

[1] MMPA amendments were included in P.L. 104-297 (§405(b)(3)), P.L. 105-18 (§2003 and §5004), P.L. 105-42 (International Dolphin Conservation Program Act), P.L. 105-277, P.L. 106-555 (Title II, Marine Mammal Rescue Assistance Act of 2000), P.L. 108-108 (§149), P.L. 108-136 (§319), and P.L. 109-479 (Title IX). For additional information, see CRS Report RL33459, *Fishery, Aquaculture, and Marine Mammal Legislation in the 109th Congress*, by Eugene H. Buck. Archived issue briefs covering legislation in previous Congresses are also available from this author.

[2] Zoos and aquariums holding marine mammals for public education and entertainment.

[3] To facilitate a candid discussion of issues, individual respondents were guaranteed they would not be identified by name. Opinions of individuals and groups may not accurately reflect the opinion of the majority. Presentation of constituent opinion in this chapter represents a sampling, and is not a quantitative assessment.

[4] The Administration's draft language of June 16, 2005, is available at [http://www.nmfs.noaa.gov/pr/pdfs/laws/mmpa_bill.pdf].

[5] These are general characterizations. There is enormous variability and crossover of membership in these groups, which often blurs the distinction among the concerns within each group. For example, marine mammal scientists may act both as objective independent analysts and serve as advocates for a specific sector.

[6] National Marine Fisheries Service, *Fisheries of the United States, 2002*, Current Fishery Statistics No. 2002 (Sept. 2003), p. 94. No revised estimate has been published more recently.

[7] National Marine Fisheries Service, Fisheries of the United States, 2005, Current Fishery Statistics No. 2005 (February 2007), p. 82. This number represents individuals employed by processors and wholesale plants. It does not include catching, transporting, or retail marketing of commercially caught fish, nor does it include jobs supported by commercial fisheries.

[8] Ex-vessel value is the money paid to the harvester for fish, shellfish, and other aquatic plants and animals, i.e., the dollar value of the harvest when it is offloaded from the boat.

[9] Supra note 7, p. 4.

[10] Id., p. 79.

[11] Some consider this action a ban or prohibition, rather than a moratorium, because it was (and is) permanent.

[12] Under the MMPA, in 16 U.S.C. §1362(13), take means "to harass, hunt, capture, or kill, or attempt to harass, hunt, capture, or kill."

[13] Although the State of Alaskabegan the process to request management authority for some marine mammal species, no state has been granted such management authority.

[14] Section 109(f)(2) defines subsistence uses as "the customary and traditional uses by rural Alaska residents of marine mammals for direct personal or family consumption as food, shelter, fuel, clothing, tools, or transportation; for the making and selling of handicraft articles out of nonedible byproducts of marine mammals taken for personal or family consumption; and for barter, or sharing for personal or family consumption."

[15] Subsequently, the MMPA Amendments of 1994 established a new regime to govern the incidental taking of marine mammals by commercial fishing operations.

[16] For more information, see CRS Report 94-751 ENR, *Marine Mammal Protection Act Amendments of 1994*, by Eugene H. Buck.

[17] Pinnipeds include seals, sea lions, and walrus.

[18] Members of TRTs "include representatives of federal agencies, each coastal state which has fisheries which interact with the species or stock, appropriate Regional Fishery Management Councils, interstate fisheries commissions, academic and scientific organizations, environmental groups, all commercial and recreational fisheries groups and gear types which incidentally take the species or stock, Alaska Native organizations or Indian tribal organizations, and others as the Secretary deems appropriate" (16 U.S.C. §1387(f)(6)(C)).

[19] For additional background on TRTs, see [http://www.nmfs.noaa.gov/pr/interactions/trt/].

[20] For individual stock assessment reports, see [http://www.nmfs.noaa.gov/pr/sars/species. htm].

[21] For additional information, see CRS Report RS22149, *Exemptions from Environmental Law for the Department of Defense: Background and Issues for Congress*, by David M. Bearden.

[22] 64 *Fed. Reg.* 24590-24592 (May 7, 1999).

[23] 65 *Fed. Reg.* 30-59 (Jan. 3, 2000).

[24] Brower v. Evans, 93 F. Supp 2d 1071, 2000 U.S. Dist. LEXIS 4624 (N.D. Cal. 2000).

[25] Brower v. Evans, 257 F. 3d 1058, 2001 U.S. App. LEXIS 16504 (9th Cir. 2001).

[26] Some scientists have attempted to define OSP for a population based on the carrying capacity for an ecosystem that may no longer exist for many reasons, both human-caused and natural.

[27] These scientists are concerned that, since most marine mammal species have suffered dramatic population decreases over the last two centuries, the true carrying capacity of the environment is unknown. In addition, they believe that carrying capacity for some species would increase if certain commercial fish harvests were curtailed and other human uses of the marine environment were modified.

[28] Maximum productivity is based on inexact population surveys subject to natural fluctuations and can be derived scientifically in several ways. Improving survey techniques with more advanced technology holds promise for improving the precision of these variables. Full realization of this potential may be dependent upon increased funding.

[29] The recovery factor accounts for uncertainty in population estimates and reproductive rates.

[30] These interests see PBR as a means of invoking the precautionary principle in marine mammal management — by which the federal government takes action to avert possible harm to marine mammals, even when the causal link between human behavior and those damages is not completely clear. For additional information on the precautionary principle, see [http://www.sehn.org/pdf/putvaluesfirst.pdf].

[31] 69 *Fed. Reg.* 43338-43345 (July 20, 2004).

[32] New methodology might include both population numbers and ecological relationships as well as review by independent scientists.

[33] "For updated stock assessments to be meaningful, this absence of sound scientific data needs to be addressed by providing for enhanced capability to conduct high priority population surveys, and studies for development of alternative population indices." Marshall Jones, Acting Deputy Director, U.S. Fish and Wildlife Service, June 29, 1999, hearing before the House Resources Subcommittee on Fisheries Conservation, Wildlife, and Oceans.

[34] Marine mammals are persistent when they discover a food source. They may habituate to acoustic devices unless these devices are quite loud, in which case the animal's hearing could become impaired. The ways and extent to which widespread use of acoustic alarms and deterrents may affect the natural ability of marine mammals to find food and use the fullextent of their foraging range is not well known, but may have unintended consequences.for example, the loss of hearing due to loud noise might increase the dependancy of marinemammals upon fishing boats and fish farms for food. In addition, AHDs could displace non-target species (such as porpoises) several miles (e.g., Retreat Passage, British Columbia).

[35] Guidelines and regulations for use of deterrents were proposed at 60 *Fed. Reg.* 22345-22348 (May 5, 1995), but NMFS never promulgated final regulations.

[36] See congressional testimony by Dr. William T. Hogarth, Assistant Administrator forFisheries, NMFS, NOAA, at [http://www.ogc.doc.gov/ogc/legreg/testimon/107f/hogarth1011.htm].

[37] Some of this research has been conducted on captive marine mammals, which may havelimited applicability to the behavior of wild, free-ranging animals.

[38] Even low-sound-output devices (e.g., "pingers") may displace animals from critical feeding habitat.

[39] For example, the physiology of different species interacting in a particular habitat, sensory processes, ecosystem implications, stock assessments, and specific behavioral characteristics/region. Studies that have been conducted are inconclusive with respect to (1) effects of a single deterrent device on multiple species inhabiting a given area (including fish); (2) audiological and physiological understanding of the marine mammal ear (and hearing thresholds); (3) impacts of both broad- and narrow-band spectra signals on the marine mammal auditory system; (4) frequency, intensity levels, and duty cycles of such devices with respect to ambient noise, vessel operations, *etc.*; and (5) acoustic behavior of the animals.

[40] Others assert that it is an extreme standard to be required to prove a negative — that an AHD does not cause harm. They claim a much more reasonable standard might be to prohibit the use of AHDs that have been shown to cause any kind of permanent damage.

[41] Some fishery practices (e.g., discarding bycatch and fish waste) invite marine mammals into close proximity with humans. In addition, an increase of fishery interactions with sperm whales in Alaskan waters appears to have coincided with the change from a "derby fishery" (where the whole fleet fished for a short period of time) to an individual fishing quota (IFQ) system (where individual fishermen choose when to fish during most of the year). The IFQ system may have enabled sperm whales to develop their skill in taking fish from fishermen. Before the change to IFQs, whales had, at most, two weeks to interact with longline fisheries and, since all vessels were fishing at the same time, not every vessel experienced problems with the whales. Now, sperm whales apparently go from boat to boat in time and space, practicing their skills most of the year.

[42] A simplified permit process might address the impacts on non-target species, and a research program could be established to assess the long-term impacts on target and non-target species from the use of AHDs.

[43] National Marine Fisheries Service. *Impacts of California Sea Lions and Pacific Harbor Seals on Salmonids and West Coast Ecosystems,* Report to Congress (Feb. 10, 1999), p. 15.

[44] Ibid., p. 15-16.

[45] The federal courts have ruled that the federal government cannot issue permits to kill an abundant animal when they know that a protected animal is also likely to be killed. See

Kokechik Fishermen's Association v. Secretary of Commerce, 839 F.2d 795 (D.C. Cir. 1988) cert denied, 488 U.S. 1004 (1989).

[46] For more information on marine protected areas, see CRS Report RS10810, *Marine Protected Areas: An Overview*, by Jeffrey A. Zinn and Eugene H. Buck.

[47] It has been suggested that some actions could be administrative (e.g., NMFS consultation on designating "essential fish habitat") such that protection of sensitive fish habitat might also acknowledge the importance of critical foraging areas for sub-adult and reproductively active female marine mammals.

[48] Section 1 of P.L. 99-625. However, FWS decided in January 2001 to halt the removal of southern sea otters from the management zone. For more information on this decision, see [http://pacific.fws.gov/news/2001/2001-23.htm].

[49] See [http://www.fws.gov/pacific/news/2005/seaotterNR.pdf].

[50] Coordinated oversight on this issue can be complicated by committee jurisdiction, since APHIS and the AWA fall under the jurisdiction of the House Committee on Agriculture and Senate Committee on Agriculture, Nutrition, and Forestry while NMFS and the MMPA areunder the jurisdiction of the House Committee on Resources and Senate Committee onCommerce, Science, and Transportation.

[51] Various aspects of this issue were recently highlighted in a five-part series, "Below the Surface," published in the *South Florida Sun-Sentinel*, May 16-19, 2004, available at [http://www.sun-sentinel.com/news/sfl-marinestorygallery,0,2119297.storygallery?coll =sfla-home-dots-utility].

[52] In August 1998, NMFS, FWS, and APHIS signed a memorandum of understanding (MOU) outlining their respective independent and collaborative roles. This MOU provides implementation strategies to ensure priority care for marine mammals, and formalizes information sharing among the agencies to promote enforcement and compliance.

[53] APHIS began the process of amending marine mammal regulations under the AWA in 1990. Subsequently, APHIS published an advanced notice of proposed rulemaking at 58 *Fed. Reg.* 39458 (July 23, 1993). Proposed regulations were published at 64 *Fed. Reg.* 8735-8755 (Feb. 23, 1999), and final regulations at 66 *Fed. Reg.* 239-257 (Jan. 3, 2001).

[54] Animal protection advocates report that APHIS employs only one veterinarian with marine mammal expertise among a staff of approximately 106 inspectors. These 106 inspectors are responsible for 8,800 licensed zoos, circuses, and trucks/airlines that transport animals. Although the AWA requires one unannounced inspection per year, animal protection groups contend that overworked inspectors visit some marine parts and aquaria only once every three years, or only after the filing of public complaints.

[55] APHIS has more than 20 years' experience in monitoring and regulating the humane care and treatment of marine mammals in captivity, employing a professional veterinary staff to inspect facilities. APHIS was given authority under the AWA to regulate warm-blooded animals, including marine mammals, for public display in the early 1970s, and first published regulations on marine mammals in 1979. APHIS resources include a National Animal Health Monitoring System, National Veterinary Services Laboratories, and a Veterinarian-in-Charge in every state.

[56] Critics suggest NMFS and FWS are already overburdened with serious problems concerning declining stocks of wild animals and a deteriorating environment.

[57] Critics cite examples where APHIS appears content to wait for facilities to fix recurring problems rather than taking more aggressive action, and where APHIS is alleged to have accepted a facility's tank measurements rather than taking independent measurements.

[58] NMFS's Marine Mammal Inventory Report now catalogs export and facility transfer notifications as required by 16 U.S.C. §1374(c)(10)(F).

[59] Comity is the legal doctrine under which countries recognize and enforce each others' legal decrees.

[60] Animal protection advocates have serious concerns regarding the ability of NMFS/FWS under the short notification regime and without public input to ensure the well-being of marine

mammals leaving this country for foreign, and often substandard, facilities. They are concerned that the brief window of notification eliminates any and all opportunity for public notification and comment and also limits the time available for the agencies to review the documentation that must accompany an export.

[61] The scientific issue is one of detailed and open record keeping, so that scientists know where animals have gone and are able to compare wild to captive mortality rates, birth rates, etc.

[62] The primary safeguard is the requirement, certified by the recipient nation's agency responsible for marine mammals, that the receiving facility meets the same criteria for holding such animals as were required of the originating U.S. facility (16 U.S.C. §1374(c)(9)). While some critics suggest that stronger regulatory criteria might be imposed by NMFS/FWS in implementing this provision, they believe such action may require congressional direction, either through a statement in committee report language or as a specific MMPA amendment.

[63] 16 U.S.C. §1374(c)(8)(B)(i)(II).

[64] For example, see [http://cerf.bc.ca/pubs/biblio/marmam_biblio.html]. Prior to the establishment of marine mammal facilities, most of what was learned about marine mammals resulted from whaling and sealing activities, rather than from field research.

[65] Some scientists suggest that a workshop of experts be convened to provide guidance on better defining what might be considered valid research on captive marine mammals, and on increasing opportunities for legitimate research access to captive marine mammals. Similar efforts have been conducted under the authority provided in 16 U.S.C. §1380.

[66] Some scientists report that research on captive animals is also constrained by economics. For example, estimates of the cost of obtaining a young healthy dolphin range from $100,000 to $150,000.

[67] However, captive marine mammals used for research often are orphaned, stranded, or disabled animals that are not physically able to be returned to the wild.

[68] Again, some believe it may be an extreme standard to be required to prove a negative.

[69] This disease is prevalent in wild animals, but has never been reported in a non-stranded captive animal.

[70] This concern may arise when private organizations, often affiliated with animal protection groups, promote the release of captive animals, as well as with some foreign public display facilities.

[71] For individual public display facilities discharging waste to a publicly owned treatment works, local municipalities enforce wastewater treatment standards and effluent discharge permits under the authority of the federal Clean Water Act. If the facility discharges directly to the environment, standards and permits under this same act are administered by the Environmental Protection Agency (EPA) or qualified states to which EPA has delegated responsibility.

[72] 66 *Fed. Reg.* 239-257 (Jan. 3, 2001).

[73] In rare instances, such as hazardous situations or removal of an animal from imminent danger, it may be in the sick or threatened animal's best interests to be transported to a quarantined location where it can be treated. Animal protection interests are concerned to ensure that cumbersome paperwork and bureaucracy do not jeopardize an animal's life in these situations.

[74] In the reverse situation, cases have been reported where receipt of a stranded wild animal or inadequate treatment of influent water has allowed pathogens from the wild to infect captive marine mammals.

[75] Necropsies are routinely performed as soon as possible, normally within hours of death. Histopathological samples are collected and a full spectrum of tests are conducted by independent laboratories outside the facility. A full report of test results is normally received within two to three weeks, with preliminary results usually available within a week.

[76] 9 C.F.R. 3.110(d).

[77] 16 U.S.C. §1374(c)(10}(H). NMFS requires, by policy, that deaths be reported within 30 days, and has announced its intent to put this policy into regulation.

[78] With few exceptions, zoos and aquaria claim to be open regarding the disposition of marine mammals within their care, with records available for public review. Animal protection advocates dispute this claim, suggesting that a majority of facilities refuse to share such information, considering it proprietary.

[79] Animal protection advocates believe that considerable incentive exists for public display facilities to provide false or incomplete information on the cause of death of marine mammals, asserting that these institutions are unlikely to provide evidence that would lead to accusations of wrongdoing, with subsequent scrutiny and possible fine.

[80] Supporters of supervised or independent necropsies contend that requirements for such might further protect facilities from liability and claims of negligence, whereas the current unsupervised necropsies may leave them unprotected.

[81] 66 *Fed. Reg.* 239-257 (Jan. 3, 2001).

[82] Alternatively, it could be required that genetically mixed offspring be neutered before release.

[83] Some scientists contend that, while encouraging breeding among groups of animals taken from the same general population may be appropriate, encouraging mixing between populations makes little sense given what is known about the movements and social isolation of many species of marine mammals.

[84] Similar efforts have been conducted under the authority provided in 16 U.S.C. §1380.

[85] Some public display interests and managers suggest that captive care and maintenance practices are constantly evolving and improving such that historic survivorship data might have limited relevance to the current situation. In addition, others suggest that survivorship is so highly variable that it would be difficult to compare populations, captive and/or wild, and come to any statistically significant conclusions.

[86] Similar efforts have been conducted under the authority provided in 16 U.S.C. §1380.

[87] 66 *Fed. Reg.* 239-257 (Jan. 3, 2001).

[88] Animal protection advocates cite several instances where dolphins and pilot whales are alleged to have been successfully released, with subsequent observation of apparently successful social integration with wild animals over a period of time.

[89] A scientific workshop might be convened to develop the protocols for conducting rehabilitation/release projects.

[90] How such facilities and programs might interact with existing marine mammal stranding networks would need to be defined. These networks along the Atlantic, Gulf of Mexico, and Pacific Coasts involve dozens of facilities that provide short-term assistance to beached and stranded marine mammals when necessaryto improve their condition sufficiently to be able to return a healthy animal to the wild.

[91] These critics suggest that veterinary examinations are unlikely to be able to pronounce captive animals disease-free, and that released animals are unlikely to be accepted easily or smoothly into the social structure of wild populations.

[92] While some coastal species may inhabit a topographically diverse physical environment, the open ocean is almost featureless. Some scientists suggest that emphasis should be placed on cleanliness, space, and behavioral responses, rather than what humans might assume constitutes a "quality" environment, since most marine mammals get their stimulation from social interaction, feeding, etc.

[93] Generally, captive holding facilities and marine mammal trainers assume responsibility for providing environmental enrichment in the form of playtime, toys, and other stimulating objects or activities. In addition, facility design criteria have changed substantially to where habitats currently under construction incorporate innovative shapes and varying rockwork for alternating surfaces, providing swim-through areas (arches and tunnels) as well as areas for rubbing and scratching.

[94] 64 *Fed. Reg.* 15918-15920 (Apr. 2, 1999). On May 30, 2002, APHIS sought comments on standards for interactive swim-with-the-dolphin programs (67 *Fed. Reg.* 37731-37732). No

final rule has yet been published. For additional background on these programs, see *Quantitative Behavioral Study of Bottlenose Dolphins in Swim-With-The-Dolphin Programsin the United States at* [http://www.nmfs.noaa.gov/pr/pdfs/health/swimwith dolphins.pdf].

[95] This document was available at [http://iucn.org/dbtw-wpd/edocs/2003-009.pdf].

[96] Some managers suggest this is due, in part, to limited funds appropriated by Congress to the various agencies, especially NMFS.

[97] These MTRPs do not collect data useful for accurately assessing the age/sex composition of the harvest, nor for establishing annual productivity. Although it might require additional agency funding, MTRPs could be restructured to obtain these data.

[98] On May 24, 1999, NMFS published an interim final rule requiring the marking and reporting of beluga whales harvested from Cook Inlet (64 *Fed. Reg.* 27925-27928).

[99] The nature of human relationships in small rural Alaskan communities makes obtaining consistently accurate data extremely difficult. Thus, the precision and accuracy of retrospective household surveys for marine mammal harvest is questioned by some critics, especially where such work has not been independently peer-reviewed. Such retrospective surveys for marine mammal harvest might be considered minimum estimates.

[100] Exemptions from reporting might be granted when or where stocks are not in decline, not listed under the ESA, and not harvested at levels exceeding 10% of the PBR level.

[101] Some critics fear that federal managers may be pressured to set PBR levels higher than the subsistence harvest levels for some Alaskan species (e.g., Pacific walrus).

[102] Many Native Alaskans, regardless of where they reside, are employed full-time with limited opportunity to continue hunting and gathering to support their traditional subsistence lifestyle and diet. Thus, the commercial marketplace may provide their only access to traditional foods, which is part of maintaining a cultural identity.

[103] Whale skin and adhering blubber.

[104] In the late 1990s, this was seen as a particular problem for the Cook Inlet beluga whale stock, which was small and had been overharvested, largely because of market hunting. A significant percentageof the Cook Inlet beluga whale stock was killed each year — between 98 and 147 animals were reportedly taken in 1996, with another 49 to 98 animals struck and lost. This stock declined almost 50% in abundance from an estimated 653 animals in 1994 to 347 animals in 1998, and its summer range contracted.

[105] In the Cook Inlet beluga whale example, Congress acted in section 3022 of P.L. 106-31 to prohibit subsistence hunting of Cook Inlet beluga whales during FY2000 to give NMFS time to take administrative action. Subsequently, NMFS conducted a status review of this stock and designated it as depleted under the MMPA (65 *Fed. Reg.* 34590-34597, May 31, 2000), but determined that listing the stockas endangered under the ESA was not warranted (65 *Fed. Reg.* 38778-38790, June 22, 2000).

[106] A "strike" means hitting a whale with a harpoon, lance, or explosive device.

[107] However, in the example of the Cook Inlet beluga whales, critics fault NMFS for relying upon the Cook Inlet Marine Mammal Council to develop some mechanism for self-regulation, which it was slow to do.

[108] Makah whaling was suspended on June 9, 2000, by the Ninth Circuit Court of Appeals (*Metcalf v. Daley*, No. 98-36135), with NMFS ordered to begin the National Environmental Policy Act (NEPA) process afresh and prepare a new environmental assessment. Subsequently, NMFS set the Makah gray whale quota at zero (65 *Fed. Reg.* 75186, Dec. 1, 2000), pending completion of the NEPA analysis. On December 20, 2002, the Ninth Circuit Court of Appeals reversed a district court ruling that upheld NMFS's issuance of a quota to the Makah in 2001 and 2002 (*Anderson v. Evans*, 314 F.3d 1006 (9th Cir. 2002)). The federal government is considering whether to request rehearing of *Anderson v. Evans*. Subject to the outcome of a possible rehearing, NMFS is preparing an environmental impact statement on the issuance of annual quotas to the Makah for the years 2003 through 2007 (68 *Fed. Reg.* 10703-10704, Mar. 6, 2003).

[109] Marine Mammal Commission, *Annual Report to Congress, 1998* (Washington, DC: Jan. 31, 1999), p. 29-32.

[110] Some of these cultures might not elect to kill whales for strictly cultural benefits if commercial trade in whale products, domestically and/or internationally, was not also permitted.

[111] Background on the federal/state conflict in Alaska over subsistence use can be found at [http://www.subsistence.adfg.state.ak.us/download/subupd00.pdf].

[112] A subsequent amendment in §5004 of P.L. 105-18 relaxed criteria that needed to be met before polar bear trophies taken in Canada prior to the 1994 MMPA amendments could be imported to the United States.

[113] Canada is the only nation inhabited by polar bears that allows sport hunting. In January 2001, an emergency interim rule halted imports of polar bears taken from Canada's M'Clintock Channel population after the previously approved harvest was found to be unsustainable (66 *Fed. Reg.* 1901-1907, Jan. 10, 2001). A final rule was adopted in October 2001 (66 *Fed. Reg.* 50843-50851, Oct. 5, 2001).

[114] For additional background on this proposal, see CRS Report RL33941, *Polar Bears: Proposed Listing Under the Endangered Species Act*, by Eugene H. Buck.

[115] Provided that these takings do not cause unmitigable damage to marine mammal populations and have no more than a negligible effect on subsistence needs.

[116] NMFS justification for not regulating this activity includes the large numbers of vessels, the lack of identified cost-effective mitigation measures, the lack of authority over international vessels to implement effective mitigation measures to decrease noise effectson marine mammals, and the economic disadvantage potentially falling on those U.S.vessels that might be required to implement costly mitigation.

[117] NMFS had interpreted this to mean a portion of a marine mammal stock whose takingwould have a negligible effect on that stock. However, the ruling in *NRDC v. Evans* (279 F. Supp. 1129 (N.D. Cal. 2003)) concluded that NMFS improperly collapsed two standards and eliminated the possibility that the two standards could serve as separate safeguards restricting the extent of takes. NMFS was directed to redefine "small numbers" as a separate standard.

[118] Measurements are obtained by attaching time-depth recorders to animals which are then exposed to the sounds. For details, consult [http://is.dal.ca/~whitelab/rwb/suction.htm]. Others have used autonomous seafloor acoustic recorders that record all sounds for as long as 22 days or until batteries fail. Using such methods, whale vocalization rates have been observed to be influenced by airgun pulses from seismic surveys.

[119] This includes scientists using noise in their research as well as scientists consulting for industries and agencies (e.g., the U.S. Navy) that release large amounts of noise into the ocean.

[120] These advocates also assert that various activities with the potential to kill, injure, and harass marine mammals are regulated inconsistently and inequitably, with commercial fishing given much more liberal treatment (e.g., liberal PBRs and use of deterrents) than anthropogenic noise (e.g., concern over course deviations and other short-term behavioral changes).

[121] See the previous section, "LargeIncidental Takes," which considers whether the MMPA should be amended to regulate these activities.

[122] Reported in "Scientific Correspondence," *Nature*, Mar. 5, 1998.

[123] For more details, see [http://www.nmfs.noaa.gov/pr/pdfs/health/stranding_bahamas2000. pdf].

[124] See [http://www.geotimes.org/jan03/NN_whales.html].

[125] See 67 *Fed. Reg.* 467121-46789 (July 16, 2002).

[126] For example, see 68 *Fed. Reg.* 50123-50124 (Aug. 20, 2003).

[127] See *Natural Resources Defense Council v. Evans*, 279 F. Supp. 2d 1129 (N.D. Cal. 2003).

[128] See 68 *Fed. Reg.* 44311 (July 28, 2003).

[129] Some scientists assert that little has been published on this topic because insufficient resources to address the problem have been provided by funding agencies. They further question, if funding were provided, whether permits from NMFS and various Institutional Animal Care and Utilization Committees mandated by the Animal Welfare Act would allow necessary research to be conducted.

[130] Other than, perhaps, taxpayer groups.

[131] Some critics allege bias and/or conflict of interest in current federal agency permitting procedures, wherein applications for highly controversial research pass quickly and quietly through the review process when forwarded by field staff within the permitting agency, while comparable proposals by non-agency researchers can take months or longer to receive action.

[132] 16 U.S.C. §1374 (c)(3)(C).

[133] Some agency managers suggest three days between receipt of a permit application and publication in the *Federal Register* is reasonable and attainable.

[134] Some, but not all, of this research may already be reviewed by institutional animal committees required by 7 U.S.C. §2143 or by animal care committees required by 42 U.S.C. §289d.

[135] Also, Guam and the Northern Mariana Islands exercise similar authority.

[136] Although the State of Alaska began the process to request management authority for some marine mammal species, no state has been granted such management authority.

[137] Some suggest a limited MMPA amendment to permit importing of "byproducts of aboriginal subsistence activities," allowing, for example, ringed seal skins from Canadian and Greenland Inuit subsistence hunters to enter U.S. markets.

[138] The main concern by the WTO appears to be that the MMPA prohibits trade in marine mammal products regardless of a species' conservation status. Thus, the United States may encounter difficulties in justifying the expansive MMPA ban on imports as necessary for responding to legitimate conservation concerns. For those species where conservation is a concern, listing under the ESA provides trade restrictions under CITES.

[139] Elements of acceptable harvest management might include a sustainable kill based on sound science with adequate animal welfare standards. However, some U.S. scientists and managers might argue that, for the United States to be able to certify that our own science meets these standards, substantial expansion of U.S. research programs might be required.

[140] National Marine Fisheries Service, *Impacts of California Sea Lions and Pacific Harbor Seals on Salmonids and West Coast Ecosystems,* Report to Congress (Feb. 10, 1999).

[141] Such adverse effects include competition for fish stocks and fecal contamination of shellfish areas near seal and sea lion haulout areas.

[142] National Marine Fisheries Service, *Impacts of California Sea Lions and Pacific Harbor Seals on Salmonids and West Coast Ecosystems,* Report to Congress (Feb. 10, 1999). p. 13-15.

[143] In addition, some of these changes may also alter conditions determiningwhere pinnipeds congregate and feed, possibly increasing predation on juvenile salmon.

[144] 16 U.S.C. §1361.

[145] In addition, international dialogue on whale conservation has occurred under the auspices of the International Whaling Commission.

[146] For more information, see CRS Report RS21157, *Multinational Species Conservation Fund,* by M. Lynne Corn and Pervaze A. Sheikh.

[147] 16 U.S.C. §1362(18).

[148] In addition, some scientists believe the current definition is meaningless and possibly counterproductive. These critics suggest that an expert panel be convened to redefine this term.

[149] For example, animal protection advocates report that a pod of perhaps 50-75 spinner dolphins in Calexico Bay, Hawaii, can be surrounded on some days by as many as 50 swimmers, 35 kayaks, and several motor-propelled boats. On other days, no more than about 20 dolphins come into the Bay, where they are pursued from early morning until late afternoon when they leave the bay. NMFS doesn't have an enforcement agent on the Big Island (where

these violations occur), and an agent from the Hawaii Department of Land and Natural Resources is responsible for responding to possible violations.

[150] Others suggest the problem is regulatory, wondering why U.S. agencies do not adopt an approach similar to that of Mexico where the number of vessels that can be in the proximity of any whale or group of whales is strictly limited and enforced. Some suggest that the revised operational guidelines for whale-watching vessels in the northeastern United States (64 *Fed. Reg.* 29270-29271, June 1, 1999) are a positive step, and that additional region- or area-specific guidelines or regulations of a similar nature should be developed.

[151] Others, however, find MMPA management inconsistent in making it illegal to swimwith wild dolphins who willingly approach humans while allowing commercial ventures to hold dolphins captive and charge humans for the chance to swim with them.

[152] Current requirements for prosecuting harassment violations are stringent, requiring a time-/date-stamped video of the incident and a court appearance by the complainant to testify against the offender.

[153] For additional information, see "Military Readiness and Environmental Exemptions" in CRS Report RL32183, *Defense Cleanup and Environmental Programs: Authorization and Appropriations for FY2004*, by David M. Bearden.

[154] For example, FWS uses MTRP (see "Reporting Subsistence Takes") while NMFS does not, and NMFS uses "incidental harassment authorization" to permit incidental taking while FWS does not.

[155] Joint regulations relating to marine mammals have only been developed for the transfer of management authority to states (50 C.F.R. Part 403).

[156] National Marine Fisheries Service, *Impacts of California Sea Lions and Pacific Harbor Seals on Salmonids and West Coast Ecosystems,* Report to Congress (Feb. 10, 1999), p. 16-17.

[157] Such a program might be authorized as an extension of the Pacific Coast Task Force provisions in 16 U.S.C. §1389.

[158] For background on this issue, see CRS Report RL31267, *Environmental Exposure to Endocrine Disruptors: What Are the Human Health Risks?* by Linda-Jo Schierow and Eugene H. Buck.

[159] For example, the Humane Society of the United States was a plaintiff in at least one lawsuit pertaining to perceived NMFS inaction on take reduction mandates in the 1994 MMPA amendments; the Center for Biological Diversity also has filed suit against NMFS for failure to convene a take reduction team.

[160] On June 29, 1999, Marshall Jones, Acting Deputy Director, U.S. Fish and Wildlife Service, testified before the House Resources Subcommittee on Fisheries Conservation, Wildlife, and Oceans that "due to competing budget needs and limited funding, the Service has been unable to fully implement provisions of certain amendments."

[161] U.S. Congress, House Committee on Resources, Subcommittee on Fisheries Conservation, Wildlife, and Oceans, *Marine Mammal Protection Act*, 107th Cong., 1st sess. (Oct. 11, 2001), Serial No. 107-65, 330 p.

[162] U.S. Congress, House Committee on Resources, Subcommittee on Fisheries Conservation, Wildlife, and Oceans, *H.R. 4781, Marine Mammal Protection Act Amendments of 2002*, 107th Cong., 2nd sess. (June 13, 2002), Serial No. 107-128, 94 p.

[163] The Pew Oceans Commission's report, *America's Living Oceans: Charting a Course for Sea Change*, was available at [http://www.pewtrusts.com/pdf/env_pew_oceans_final_report.pdf].

[164] The U.S. Commission on Ocean Policy's preliminary report, *Preliminary Report of the U.S. Commission on Ocean Policy,* was available at [http://oceancommission.gov/documents/prelimreport/welcome.html].

In: Marine Mammal Protection Issues
Editors: Derek L. Caruana

ISBN: 978-1-60741-540-4
© 2010 Nova Science Publishers, Inc.

Chapter 2

NATIONAL MARINE FISHERIES SERVICE: IMPROVED ECONOMIC ANALYSIS AND EVALUATION STRATEGIES NEEDED FOR PROPOSED CHANGES TO ATLANTIC LARGE WHALE PROTECTION PLAN

United States Government Accountability Office

WHY GAO DID THIS STUDY

The National Marine Fisheries Service (NMFS) developed the Atlantic Large Whale Take Reduction (ALWTR) plan to protect endangered large whales from entanglements in commercial fishing gear, which can cause injury or death. Because whales continued to die after the ALWTR plan went into effect, NMFS proposed revisions in 2005. GAO was asked to review these proposed revisions, including (1) their scientific basis and uncertainties regarding their effectiveness, (2) NMFS's plans to address concerns about the feasibility of implementing them, (3) the extent to which NMFS fully assessed the costs to the fishing industry and impacts on fishing communities, and (4) the extent to which NMFS developed strategies for fully evaluating their effectiveness. GAO reviewed the proposed changes to the ALWTR plan and obtained the views of NMFS officials, industry representatives, scientists, and conservationists.

WHAT GAO RECOMMENDS

GAO recommends that NMFS revise its economic analysis to present a range of possible costs, expand its proposed gear-marking requirements, and develop a strategy to assess industry compliance. The agency reviewed a draft of this chapter and did not agree to revise its economic analysis or expand gear markings but did agree to develop a strategy to assess industry compliance.

WHAT GAO FOUND

NMFS used scientific data on whale entanglements, scarification, and sightings as support for its proposed changes to the ALWTR plan. These data indicate that right and humpback whales are being injured and killed by entanglements in commercial fishing gear at a rate that limits the species' ability to recover. One of the key proposed changes to the ALWTR plan involves replacing floating groundline, which forms arcs in the water that can entangle whales, with sinking groundline that lies on the ocean bottom. While there is a consensus among whale experts that using sinking groundline will reduce risks to whales, uncertainties remain regarding how many fewer serious injuries and mortalities will occur as a result of this requirement.

NMFS has not yet resolved implementation issues associated with using sinking groundline in rocky bottom areas, particularly off the coast of Maine. While NMFS believes that it is operationally feasible to use sinking groundline in all areas, it recognizes that fishermen may have to modify their fishing practices to use this type of gear effectively. Maine lobster industry representatives told GAO that fishermen who operate in rocky bottom areas will not be able to use sinking groundline because it will wear away and create safety hazards if the line snaps when it is hauled.

NMFS's economic assessment of the costs of the proposed gear modifications did not reflect the significant uncertainties associated with the assessment, and the extent to which these costs to the fishing industry could be higher or lower than reported is unclear. Because NMFS lacked verifiable data for some of the key cost variables, it used estimates and assumptions that introduced a significant amount of uncertainty into the cost calculations, which the agency acknowledged. However, instead of presenting a range of costs to account for these uncertainties, NMFS produced a single estimate of compliance costs—about $14 million annually. Moreover, because it lacked key data on fishermen's ability to absorb these costs without going out of business, NMFS could not fully assess the impacts that the cost of gear modifications would have on fishing communities. For example, without knowing which specific fishermen would go out of business, NMFS could not determine the impact lost jobs would have on the communities in which they lived.

NMFS has not developed strategies for fully evaluating the effectiveness of the proposed regulatory changes. Specifically, NMFS's gear-marking requirements may not be adequate for effectively assessing future whale entanglements because they do not include comprehensive markings that researchers could use to assess the type of rope involved in entanglements. Additionally, NMFS does not yet have a strategy to monitor the level of industry compliance and therefore lacks a means to determine whether any future entanglements are due to industry noncompliance with the regulatory requirements or the ineffectiveness of the gear modifications.

ABBREVIATIONS

ALWTR	Atlantic Large Whale Take Reduction
DAM	Dynamic Area Management
DEIS	Draft Environmental Impact Statement
ESA	Endangered Species Act
FEIS	Final Environmental Impact Statement
MLA	Maine Lobstermen's Association
MMPA	Marine Mammal Protection Act
NMFS	National Marine Fisheries Service
NOAA	National Oceanic and Atmospheric Administration
SAM	Seasonal Area Management

July 20, 2007

The Honorable Olympia J. Snowe
Ranking Member, Subcommittee on Oceans,
Atmosphere, Fisheries and Coast Guard
Committee on Commerce, Science, and Transportation
United States Senate

Dear Senator Snowe:

Despite regulatory actions designed to ensure their safety and survival, endangered large Atlantic whales continue to become entangled in commercial fishing gear, sometimes resulting in death or severe injury. Right, humpback, and fin whales are three species of Atlantic large whales that are protected under the Endangered Species Act (ESA) and Marine Mammal Protection Act (MMPA), under the administration of the National Marine Fisheries Service (NMFS).[1] NMFS is particularly concerned about the North Atlantic right whale because scientists estimate that there are only about 300 of these whales in existence. NMFS has determined that with a population reduced to such a low number, the death or serious injury of even one right whale from human-related causes, such as fishing gear entanglement, would limit the ability of the species to recover.

Atlantic large whales are at risk of entanglement in fishing gear because they feed, travel, and breed in areas where commercial fishermen leave traps and gillnets.[2] Fishermen set lobster and other traps either singly, or in strings of multiple traps linked together with rope known as groundline, as shown in figure 1. A buoy at the surface, which fishermen use to locate their gear, is connected to a vertical rope linked to the traps. Fishermen use the vertical rope to haul traps into their boats. Gillnet fisheries, which catch fish such as sharks and groundfish, use some of the same gear components, but use nets instead of traps.[3]

When whales become entangled in fishing gear, they can sometimes free themselves without serious injury. However, in other cases, entanglement can impede the whale's normal breathing and movement, causing it to drown. Even if the whale is eventually able to break free, part of the gear may remain attached to its body, sometimes making it more difficult to breathe, feed, and travel, and possibly leading to an early death.

In 1997, under the MMPA, NMFS developed the Atlantic Large Whale Take Reduction (ALWTR) Plan to reduce the risk of serious injury and mortality to right, humpback, and fin whales from entanglement in commercial fishing gear.[4] This plan included several gear modifications that apply to lobster and certain gillnet fisheries—such as prohibiting floating vertical line at the surface—as well as season-specific requirements that are in effect when whales are expected in certain areas. Due to the continued serious injury and mortality of large whales after the ALWTR plan was implemented, NMFS established additional measures. For example, in 2002, NMFS established measures (1) restricting commercial fishing gear in areas where right whales are known to feed and (2) allowing the agency to temporarily restrict or prohibit gear in specific areas of the north Atlantic if three or more right whales were observed within 75 square nautical miles.

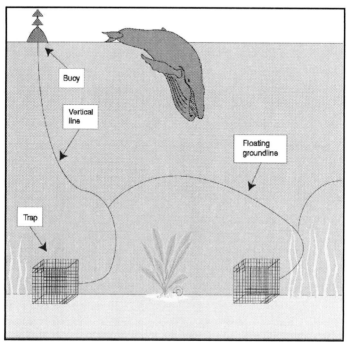

Source: GAO.

Figure 1. Commercial Gear Configurations for Trap Fisheries.

Despite NMFS's efforts, whale entanglements and deaths continued. At the end of 2002, NMFS determined, after an independent peer review, that a right whale had been entangled in gear consistent with U.S. fishing gear. Due to this and other fatal and nonfatal entanglements of right, humpback and fin whales, NMFS filed a notice of intent in the June 30, 2003, *Federal Register* that it planned to prepare an environmental impact statement to analyze the impacts of revising the ALWTR plan and stated that it would hold meetings with stakeholders to collect information on strategies to reduce whale entanglements. Between 2003 and 2004, after the stakeholder meetings, the agency developed proposed modifications to the ALWTR plan and conducted an analysis on the effects these modifications would have on whales, the fishing industry, and fishing communities. In February 2005, the agency issued a draft environmental impact statement (DEIS) that identified six alternative sets of proposed modifications to the existing ALWTR plan.[5] NMFS designated two of these as "preferred" alternatives with the goal of selecting one in the final environmental impact statement. The preferred alternatives outlined a broader approach to whale protection by incorporating additional fisheries into the ALWTR plan and requiring year-round and seasonal gear modifications in the North Atlantic. One of the key proposed changes requires fishermen to replace floating groundline, which creates arcs in the water that can entangle whales, with sinking groundline, which lies on the ocean bottom.[6] However, there are concerns that the cost of the gear modifications, particularly sinking groundline, may threaten the livelihood of fishermen, especially lobstermen. In the DEIS, NMFS estimated that the total cost to the fishing industry would be about $14 million annually and that the lobster industry would incur more than $12.8 million of these projected costs.

In June 2005, NMFS published a proposed rule to amend the regulations implementing the ALWTR plan.[7] In February 2007, after an interagency review, NMFS withdrew the rule. According to a NMFS official, the interagency review raised concerns that NMFS had not fully addressed issues raised by the state of Maine and the Maine lobster industry, such as which areas along the Maine coast should be exempt from the proposed gear modifications. NMFS is currently reevaluating the proposed regulation to determine if any revisions are needed. The agency hopes to complete its review and have a final regulation in place by year-end 2007. In the meantime, the current regulations remain in effect, and endangered large whales continue to be at risk of entanglement in commercial fishing gear.

Since NMFS has not issued a final environmental impact statement or regulation, you asked us to review the proposed changes to the ALWTR plan outlined in the DEIS. Specifically, you asked us to (1) describe the scientific basis for the proposed changes to the ALWTR plan and the extent to which uncertainties exist regarding how effectively they will protect large whales; (2) describe how the agency plans to address implementation issues, particularly in the rocky bottom areas of the North Atlantic coast; (3) evaluate the extent to which NMFS fully assessed costs to the fishing industry and the economic impacts on fishing communities; and (4) evaluate the extent to which NMFS has developed strategies for fully assessing the effectiveness of and industry compliance with the proposed changes.

To address our objectives, we reviewed the DEIS, public comments on the DEIS, and scientific literature on right, humpback, and fin whales. We also obtained the views of a wide range of stakeholders on the proposed changes to the ALWTR plan, such as marine mammal scientists, including those at the

Woods Hole Oceanographic Institution and the Provincetown Center for Coastal Studies;[8] federal regulators, including officials at NMFS's Northeast Regional Office who participated in developing the proposed changes to the plan; state fisheries management officials in Maine and Massachusetts; industry groups, including the Maine Lobstermen's Association; a conservation group, the Humane Society of the United States; and the Marine Mammal Commission, an independent U.S. agency responsible for providing oversight of the marine mammal conservation policies and programs carried out by federal regulatory agencies. We also met with officials from Industrial Economics Inc., who conducted the economic analysis for NMFS that was included in the DEIS. Finally, we reviewed documentation of federal and state compliance efforts related to the current ALWTR plan. A more detailed description of our scope and methodology is presented in appendix I. We performed our work between August 2006 and June 2007 in accordance with generally accepted government auditing standards.

RESULTS IN BRIEF

NMFS based its proposed changes to the ALWTR plan on scientific research that indicated that whales are becoming entangled in commercial fishing gear and that sinking groundline will almost certainly reduce entanglements; however, the agency cannot determine the overall extent to which the proposed gear modifications will reduce serious injury or mortality to whales. To support the need for the proposed changes to the ALWTR plan, NMFS used its scientific stock assessments and entanglement reports, which showed that—despite current regulatory measures—right and humpback whales are being seriously injured or killed by entanglements in commercial fishing gear at a rate that limits the species' ability to recover. NMFS also relied on scientific research that showed that about three-quarters of the right whale population and one-half of the humpback whale population had scars caused by entanglement with commercial fishing gear. NMFS developed the specific proposed gear modifications based, in part, on a study of gear found on entangled right and humpback whales that indicated that all parts of commercial fishing gear create a risk of entanglement for these whales. However, the study did not provide information regarding the extent to which each component of fishing gear poses a risk to whales. Therefore, NMFS could not estimate how many fewer serious injuries and mortalities will occur as a result of its proposed changes. While scientists believe that sinking groundline—one of the key features of the proposal—will reduce risks to whales, they are uncertain if it will eliminate all serious injuries or mortalities from entanglements in groundline. In addition, the study of gear found on entangled right and humpback whales indicated that other parts of the gear, including vertical line, also posed an entanglement risk. Although NMFS has taken some actions to mitigate this risk, such as implementing weak link requirements, the agency acknowledges that more needs to be done, and it plans to further address vertical line in the future.

NMFS has not resolved challenges associated with implementing the proposed fishing gear modifications in the rocky bottom areas of the North Atlantic coast. NMFS maintains that it is operationally feasible to use sinking groundline in all areas, but the agency told us that fishermen may have to modify their fishing practices. For example, fishermen may need to modify the way they retrieve their gear so that sinking groundline does not become caught on rocks, causing gear loss. However, Maine lobster fishermen contend that it is not operationally feasible for them to use sinking groundline in rocky bottom areas because the rocks will cause abrasion—wearing away or weakening the rope—which could require them to replace their rope too frequently or cause gear loss. Fishermen are also concerned that sinking groundline poses safety risks to them. For example, if sinking groundline abrades along the rocky bottom and breaks when fishermen retrieve their gear, the line could strike and injure them. A NMFS official maintained that fishermen need to be vigilant about the condition of their rope—whether it is floating groundline or sinking groundline—and replace it, as needed, to reduce the risk of injury and avoid gear loss. In January 2007, the Maine Department of Marine Resources submitted a proposal to NMFS that would allow fishermen to use "low-profile" groundline—a rope that floats on average about 3 feet above the ocean bottom—as an alternative to the use of sinking groundline along rocky bottom areas of Maine's coast. The state believes low-profile groundline will both benefit the lobster industry and protect whales. NMFS and the scientists with whom we spoke are unsure if low-profile groundline will reduce the risk of whale entanglement because it could form an arc similar to that of floating groundline creating an entanglement risk for large whales.

NMFS's economic assessment of the proposed fishing gear modifications did not (1) adequately represent the uncertainties of its cost estimates, which could result in higher or lower costs to the fishing industry than reported in the DEIS and (2) fully assess the impacts of the increased costs on affected fishing communities. NMFS included key variables, such as the cost of rope replacement and expected increases in gear loss, in its estimate of the costs of the proposed changes on the fishing industry. However, NMFS did not have verifiable data to estimate the costs of these variables. For example, NMFS's estimates of the costs of gear loss were based on expert opinions, not on data that had been verified through field testing. The use of estimates and lack of verifiable data introduced a significant amount of uncertainty into NMFS's calculations of the cost of the proposed gear modifications on fishermen. Although the agency acknowledged these uncertainties in the DEIS, it produced a single estimate of compliance costs—about $14 million annually, most of which would be incurred by the lobster industry—rather than a range of possible costs. Presenting a range of costs would have better represented the significant uncertainty that exists in NMFS's estimate and would have better demonstrated the extent to which total costs to fishermen and the fishing industry could be different than what NMFS estimated. In addition, because NMFS did not have data on fishermen's ability to absorb the costs of the proposed gear modifications, the agency used revenue estimates and made arbitrary assumptions to estimate the number of fisherman that would go out of business because of the increased costs. However, because fishermen's revenues and their ability to absorb additional costs could be noticeably different than what NMFS assumed, the number of fisherman that would go out of business could be lower or higher than NMFS estimated. Furthermore, because NMFS lacked information about which specific fishermen, living in which communities, would go out of business, it could not identify exactly which communities would lose jobs or determine the impact any lost jobs and income would have on these fishing communities.

NMFS has not developed strategies for fully evaluating the effectiveness of the proposed regulatory changes. Specifically, NMFS could require comprehensive markings on commercial fishing gear that would enable researchers to assess the type of rope involved in entanglements. Although NMFS's proposed modifications to the ALWTR plan include new gear marking requirements—such as marking vertical lines—it has not proposed marking sinking groundline because it believes that the use of sinking groundline will be completely effective in protecting whales. However, scientists with whom we spoke, including NMFS's scientists, said that while they believe sinking groundline will reduce risk of whale entanglements, they also believe its success cannot be guaranteed; and therefore, it should be marked so that its performance can be evaluated. To assess the effectiveness of its proposed regulatory requirements, NMFS also needs to be able to determine whether any future entanglements are due to noncompliance by industry with the regulatory requirements or the ineffectiveness of the gear modifications. However, NMFS has not yet developed a strategy for monitoring the level of industry compliance.

Given the need to fully disclose the potential cost burden on fishermen and to assess the proposed measures to protect endangered large whales, we are recommending that when NMFS finalizes the proposed changes to the ALWTR plan it revises its economic analysis to present a range of possible costs, expands its proposed gear-marking requirements, and develops a strategy to assess industry compliance. In commenting on a draft of the report, the National Oceanic and Atmospheric Administration (NOAA) did not agree with our first two recommendations but did agree to develop a strategy for assessing industry compliance. NOAA believes that the uncertainty of the data was adequately represented in the DEIS and therefore did not agree that the agency needs to present a range of possible costs in its final economic analysis. Nonetheless, NOAA said that it is planning to clarify the variations and uncertainties within its analysis in the Final Environmental Impact Statement. With regard to our recommendation on markings for sinking groundline and gear in exempted areas, NOAA stated that such markings are not feasible or practical at this time. It is unclear to us why NOAA would make such a statement given that in the DEIS, NMFS has proposed similar marking requirements for vertical line. Although NOAA agreed with our recommendation to develop a strategy for assessing industry compliance with the gear modification requirements, it did not believe that the recommendation could be implemented before NMFS finalizes the proposed regulations. We believe that if NOAA is unable to complete its strategy prior to finalizing its proposed regulations, the strategy should be in place by the effective date of the final regulations. The full text of NOAA's comments and our responses appears in appendix II.

BACKGROUND

Right, humpback, and fin whales were hunted by commercial whalers. The right whale, in particular, was targeted by whalers because it is a slow-moving animal that floats when it is killed, due to its high blubber content. Accordingly, whalers gave the right whale its name because it was the "right" whale to hunt. In 1949, the International Convention for the Regulation of

Whaling protected right whales from commercial whaling. In 1970, the species was listed as endangered under the Endangered Species Conservation Act, the precursor to the ESA. Right whales were subsequently listed as endangered under the ESA in 1973. Despite several decades of conservation efforts, the right whale has struggled to recover due to low reproductive rates and accidental human-caused mortality. The North Atlantic right whale is among the most endangered large whale species in the world. A 1999 study estimated that the species will be extinct within 200 years if mortality rates continue.[9] Humpback and fin whales were hunted for oil, meat, and materials for utilitarian products (e.g., corset stays, umbrella ribs, buggy whips, etc.) until the 20th century. The International Whaling Commission banned commercial whaling of North Atlantic humpback whales in 1955. Commercial whaling of the fin whale was banned in the North Atlantic in 1987.[10] Both humpback and fin whales have been listed as endangered under the ESA since its passage in 1973.

Atlantic large whales are at risk for entanglement in commercial fishing gear when they are traveling, feeding, and breeding. For example, right whales feed with their mouths open for extended periods of time using their baleen—a substance that grows in comb-like rows from the upper jaws of toothless whales—to filter plankton from seawater. Much about the movements and habitats of right whales remains unknown. However, it is generally thought that some right whales winter in the lower latitudes—off the southeast U.S. Atlantic coast, where calving takes place—then migrate to higher latitudes, near Massachusetts and Maine for the summer, following concentrations of copepods, their principal food source.[11] Right whales primarily use the mid-Atlantic region to migrate to and from the calving grounds in the south. Like right whales, humpback whales also feed off the coasts of Massachusetts and Maine, however, they winter farther south. Humpback whales employ a variety of feeding techniques that differ from right whale feeding techniques. For example, one way that humpback whales feed is by lunging into a patch of small fish with their mouth wide-open for a short period of time. Like right and humpback whales, scientists believe that fin whales use northern waters primarily for feeding and southern waters primarily for calving. Fin whales also engage in lunge feeding.

Under the MMPA, NMFS must develop a plan to protect Atlantic large whales from entanglements that cause serious injury or mortality.[12] The MMPA was enacted in 1972 to provide protection for all marine mammals. Section 118, enacted in the 1994 amendments to the MMPA, specifically outlines a process for reducing serious injury and mortality incidental to commercial fishing operations.[13] Under that process, if NMFS determines that a species' ability to recover has become diminished by commercial fishing activities, the agency must develop and implement a plan—known as a take reduction plan[14]—to reduce serious injury and mortality to the species. The MMPA requires a take reduction team to be involved in developing a take reduction plan. Members of the team are required to have either biological/conservation expertise relevant to the marine mammal species addressed in the take reduction plan or the fishing practices that result in the incidental mortality and serious injury of the species. Team members must include representatives of federal agencies, state agencies, Regional Fishery Management Councils,[15] interstate fishery commissions, academic and scientific organizations, environmental groups, and fishery groups that use gear that could harm the species.

The immediate goal of a take reduction plan is to reduce, within 6 months, mortality and serious injury below the potential biological removal level—meaning the maximum number of human-related mortalities that can occur annually without limiting the species' ability to recover.[16] The long-term goal of a take reduction plan is to, within 5 years, reduce fishery-related mortality and serious injury to insignificant levels approaching zero.[17] The take reduction plan must include recommended regulatory and voluntary measures aimed at reducing mortality and serious injury, such as requiring the use of alternative commercial fishing gear or techniques.

The current ALWTR plan, originally developed in 1997, includes both universal gear modifications that apply to all lobster traps and anchored gillnets as well as area- and season-specific requirements. The universal requirements prohibit floating vertical line at the surface, require gear to be hauled out of the water at least once every 30 days, and encourage fishermen to maintain knot-free vertical lines. In particular areas, such as Cape Cod Bay, fishermen are required to use sinking groundline, which poses less of an entanglement risk because it sinks to the ocean floor rather than creating loops in the water. Fishermen in certain areas are also required to attach weak links—devices that are designed to break if a particular amount of pressure is applied—to their vertical lines or gillnet panels and place marks on their gear so researchers may be able to identify the fishery involved and the location where the gear was set if it is later recovered from an entangled whale. In addition, certain restricted areas are closed to lobster trap fishing or anchored gillnetting during particular seasons when whales are likely to be in the area. When these areas are open, fishermen are limited to using gear that meets particular requirements, such as weak links.

While NMFS has developed the ALWTR plan pursuant to its responsibilities under the MMPA, NMFS also has responsibilities under the ESA. The ESA directs all federal agencies to utilize their authorities to conserve threatened and endangered species. In addition, such species and their habitats must be protected against adverse effects of federal activities such as operating hydroelectric dams, thinning vegetation to prevent wildfires, and—as in this case—permitting fishing, so that the continued existence of protected species is not jeopardized. The right, humpback, and fin whale species are all listed as endangered under the ESA. Section 7 of the ESA directs federal agencies that are taking actions that may affect protected species—referred to as action agencies—to initiate a "consultation" to assess the impacts of their actions on threatened and endangered species. Federal action agencies consult with either NMFS or the U.S. Fish and Wildlife Service within the Department of the Interior, depending on which species their actions may affect.[18] These agencies are referred to as the consulting agencies. For example, because NMFS regulates commercial fishing and the activities of the fishing industry have seriously injured or killed endangered whales, NMFS must consult on its proposed fishery regulation that may affect endangered whales. Consequently, in this case, NMFS acts as both the action agency and the consulting agency. Action agencies submit biological assessments to the consulting agencies that discuss their proposed activities and their likely effects on protected species and their habitat. The consulting agency completes the formal consultation process by issuing a biological opinion. If the consulting agency concludes that the proposed activities are likely to jeopardize the species' continued existence or adversely modify its habitat, the biological opinion will include reasonable and prudent alternatives that are necessary or appropriate to minimize impacts to protected species. If any "take" of a species is expected to occur, the biological opinion also must contain terms and conditions designed to reduce take and address adverse modification of designated critical habitat. For example, NMFS has prepared biological opinions to assess the impact of continuing to permit the multispecies, spiny dogfish, monkfish, and lobster fisheries on protected marine species.[19] In the most recent opinion, NMFS identified the fishing gear modifications contained in the ALWTR plan as a reasonable and prudent alternative to protect right whales from fishing gear entanglements.

In 2000, after new whale entanglements caused serious injuries to right whales, as well as at least one right whale fatality in gillnet gear, NMFS reinitiated a section 7 consultation for the multispecies, spiny dogfish, monkfish, and lobster fisheries. NMFS's biological opinion found that its administration of these fisheries was likely to jeopardize the continued existence of the right whale. Consequently, NMFS developed the Seasonal Area Management (SAM) and Dynamic Area Management (DAM) programs as reasonable and prudent alternatives to avoid further jeopardizing the existence of the right whale. The SAM program imposes seasonal restrictions on lobster and gillnet fishermen to protect predictable aggregations of right whales that annually feed in waters north and east of Cape Cod. Gear set in the SAM zone during designated times must be low-risk gear, which is defined as gear that is highly unlikely to cause death or serious injury to large whales. For example, lobster and gillnet fishermen are prohibited from using floating groundline in the western part of the SAM area from March 1 to April 30 and in the eastern part of the SAM area from May 1 to July 31, when whales are expected to be in the area. The DAM program, on the other hand, requires temporary gear restrictions in areas that experience an unexpected aggregation of right whales. If three or more right whales are spotted within 75 square nautical miles, NMFS can restrict fishing by taking one or all of the following actions: (1) requiring lobster and gillnet fishermen to remove their gear and prohibiting them from setting additional gear within the area, (2) limiting the type of gear that can be used in the area, or (3) encouraging fishermen to voluntarily stop fishing and remove their gear from the area. DAM zone restrictions remain in effect for 15 days and can be extended if three right whales continue to be sighted in the area within 75 nautical miles of each other.

Because whale entanglements that led to serious injury or mortality continued to occur after the SAM and DAM programs went into effect, in 2003, NMFS began a process of revising the ALWTR plan to require additional fishing gear modifications that apply to trap and gillnet fisheries throughout the U.S. Atlantic coast. These fisheries were selected because gear associated with them was found on entangled whales. In February 2005, NMFS issued a draft environmental impact statement under the National Environmental Policy Act that outlined its proposed regulatory changes to the ALWTR plan and the associated costs and impacts to those affected by the regulation. The DEIS identified six regulatory alternatives, two of which were identified as preferred alternatives. Some of the elements of the proposed changes were to (1) replace floating groundline with sinking groundline, (2) alter the requirements for weak links, and (3) change the gear marking requirements. Regarding weak links, NMFS proposed that lobster and other trap fisheries in some areas be required to use weak links of a lower breaking strength—making it easier for whales to break them—and that gillnet fisheries in some areas be required to use more weak links per net panel than called for in the current requirements. Regarding gear marking, NMFS proposed expanding the frequency of gear marking to one 4-inch mark every 60 feet on the vertical line, among other things.

NMFS also proposed applying these gear modifications more broadly than previous regulations. First, NMFS proposed incorporating additional trap and gillnet fisheries in to the ALWTR plan because these fisheries also use gear that poses a risk to whales.[20] Second, NMFS proposed year-round gear modifications in the North Atlantic, because whales are always present there, and seasonal gear modifications in the Mid-Atlantic and the South Atlantic regions at times when right, humpback, and fin whales sightings primarily occur.

In anticipation of increased gear costs to fishermen as a result of the proposed gear modifications, NMFS and nonprofit organizations have provided funding for fishermen to make initial replacements of floating groundline with sinking groundline. NMFS officials told us the agency recently funded a $600,000 rope buyback and recycling program for the Mid-Atlantic trap fishermen. NMFS officials also told us that the agency recently provided $2 million to the Gulf of Maine Lobster Foundation to fund a rope buyback program to assist Maine lobster fishermen. The foundation began disbursing the funds to fishermen in May 2007. In addition, NMFS officials told us the agency provided $660,000 to the International Fund for Animal Welfare, which matched the federal funding, to finance a Massachusetts rope buyback and recycling program for the lobster industry.

While fishing gear entanglement is a significant source of risk for Atlantic large whales, so are collisions with ships. For example, from 2000 to 2004, NMFS reported that one right whale and 0.6 humpback whale serious injuries or mortalities per year were attributable to collisions with ships in U.S. waters.[21] NMFS has proposed a regulation to reduce the risk of ship strikes to North Atlantic right whales, which would restrict ship speed along certain areas of the east coast during certain times of the year. NMFS expects to issue the regulation in 2007. In addition to this regulation, NMFS has also recommended changes to shipping routes off four major ports where high densities of ships and right whales overlap.

NMFS BASED PROPOSED GEAR MODIFICATIONS ON SCIENTIFIC RESEARCH, BUT IT CANNOT ESTIMATE THE EXTENT TO WHICH RISKS TO WHALES WILL BE REDUCED

Based on its scientific stock assessments of whale populations, NMFS determined that right and humpback whales are being seriously injured or killed at a rate that limits the species' ability to recover. NMFS also analyzed scientific data on whale entanglements, scarification caused by entanglement, and sightings, which supported the need to propose changes to the ALWTR plan. These data indicate that whales travel widely up and down the Atlantic coast and encounter and become entangled in commercial fishing gear. NMFS then developed the specific proposed gear modifications based, in part, on a study of the gear involved in entanglements of right and humpback whales that indicated that all parts of commercial fishing gear pose a risk to whales. While there is general agreement among scientists, conservationists, federal and state regulators, and industry groups that requiring certain commercial fisheries to use sinking groundline—one of the key features of NMFS's proposed modifications to the ALWTR plan—will reduce risks to whales, uncertainties remain regarding how many fewer serious injuries and mortalities will occur. There is also uncertainty over whether other proposed changes to the ALWTR plan will effectively protect large whales.

NMFS Based Its Proposed Gear Modifications on Scientific Studies of Whale Entanglement, Scarification, and Sightings

To support the need to propose changes to the ALWTR plan, NMFS used its annual stock assessment reports of endangered large whale populations and entanglement reports, which showed that—despite current regulatory measures—right and humpback whales were being seriously injured or killed by entanglements in commercial gear at a rate that limits the species' ability to recover to their maximum sustainable population.[22] In the 2003 stock assessment report, the agency determined—based on the size of the right whale population—that the maximum annual number of human-related mortalities that can occur without limiting the species' ability to recover is zero.[23] However, this stock assessment report showed that from 1997 to 2001, there were about 1.2 documented serious injuries and mortalities annually to right whales from fishing-gear entanglements.[24] The 2003 stock assessment report also indicated that humpback whales were being seriously injured or killed from fishing-gear entanglements at a rate that limits the species' ability to recover. The most recent stock assessment report (2006) indicates that right and humpback continue to be seriously injured or killed from fishing-gear entanglements at a rate that limits their ability to recover.[25] In contrast, NMFS determined that fin whales are not being seriously injured or killed at a rate that limits their ability to recover based on their population size and the number of serious injuries and mortalities that occur annually. Table 1 shows the data that NMFS used to assess the ability of the three species to recover based on their population size and the number of annual serious injuries and mortalities from entanglements.

Table 1. Number of Injuries and Mortalities to Large Whale Species and Impact on Their Ability to Recover.

Large whale species	Estimated population size	Average number of serious injuries and mortalities due to entanglement annually (1997-2001)[a]	Maximum number to deaths before limiting species' ability to recover
Right	291	1.2	0
Humpback	647 to 902	2.2	1.3
Fin	2,362 to 2,814	0.6	4.7

Source: NMFS data.

[a] These data include whales found in Canadian waters.

However, NMFS's annual stock assessment reports are likely to understate the full extent of whale entanglements in commercial fishing gear, as the reports only include confirmed entanglements in commercial fishing gear that have caused serious injury or mortality to whales. Additional serious entanglements may occur, but either because researchers do not recover the corpses or there is not enough evidence to determine that entanglement in commercial fishing gear caused the whales' deaths, these incidents are not captured in the stock assessment reports. A NMFS scientist with whom we spoke believes that it is likely that the agency documents only a small to modest fraction of large whale entanglements that result in serious injury or mortality. Although NMFS's stock assessment reports include data on seriously injured or dead whales found in Canadian waters, whether these whales were entangled in U.S. or Canadian gear is generally not known.[26]

In addition to the serious injuries and mortalities from entanglements documented in NMFS's stock assessment reports, NMFS also used information from scarification studies developed by various scientific institutions to demonstrate a need to revise the ALWTR plan. These studies analyzed the rate of scarring on large whales due to entanglement in fishing rope—thereby identifying the percentage of the right and humpback whale populations that experience entanglement. For these studies researchers identified individual whales using a photographic database and determined the percent that have physical evidence indicative of entanglement.[27] For example, in a 2005 report, researchers from the New England Aquarium found that approximately 75 percent of right whales had scars indicating that they had survived an entanglement in fishing rope.[28] Similarly, a 2004 report by scientists at the Provincetown Center for Coastal Studies found that approximately half of the humpback whale population also had entanglement scars.[29] However, according to a scientist with whom we spoke, these scarification studies may actually underestimate the percentage of whales that have experienced entanglement because whales that die of entanglement may not be found; researchers only count scars that they believe, based on their professional judgment, are highly likely to be from entanglement in fishing gear; and some scars may fade over time.

To determine the specific gear-modification requirements to be included in the revised ALWTR plan, NMFS relied, in part, on a study of the fishing gear found on entangled right and humpback whales conducted by NMFS researchers and gear specialists as well as researchers from the Provincetown Center for Coastal Studies and the New England Aquarium.[30] This study found that any fishing rope from trap and gillnet fisheries presents an entanglement risk to large whales because all parts of the rope, such as vertical line and groundline, have been found on entangled whales.[31]

To determine when and where to implement the proposed gear modifications, NMFS used data from the North Atlantic Right Whale Sightings Database, supplemented by additional data on humpback and fin whale sightings.[32] Using these data, researchers can identify where large whales are at risk of entanglement based on where they congregate during certain times of the year. For example, NMFS determined that right and humpback whales are sighted in the South Atlantic region from late November through early April, but are typically not present there the rest of the year. NMFS acknowledges that large whales can be found throughout the year in the Mid-Atlantic but notes that sightings occur primarily between September and May. As a result, in its preferred alternatives, the agency proposed seasonal, as opposed to year-round, gear modifications in the Mid- and South Atlantic. NMFS also used the sightings data to modify the exempted areas—those areas where commercial fishermen are not subject to the gear modifications outlined in the ALWTR plan because whales rarely, if ever, venture there.

Uncertainties Exist Regarding the Extent to Which the Proposed Gear Modifications Will Protect Large Whales

There is general agreement among scientists, conservationists, federal and state regulators, and industry groups that requiring the use of sinking groundline will reduce risks to whales. However, uncertainties remain regarding how many fewer serious injuries and mortalities will occur. NMFS was unable to quantify how much the risk of whale entanglement will be reduced by sinking groundline because researchers cannot quantify the extent to which each component of fishing gear poses a risk to whales. In addition, the scientists with whom we spoke stated that the proposed modifications to the ALWTR plan will not eliminate all entanglements because NMFS has not fully addressed the risks posed by vertical line. Although NMFS has taken some actions to mitigate the risk associated with vertical line, the agency recognizes that more needs to be done because whales continue to become entangled in this line. The agency stated that it will further address vertical line after conducting additional research and consulting with the ALWTR Team.

The scientists and conservationists with whom we spoke or who provided written comments to NMFS on the DEIS are also uncertain about the effectiveness of other aspects of the proposed changes to the ALWTR plan. Specifically, they were uncertain about whether the use of weak links will reduce risks to whales because whales have been found entangled in fishing rope that had weak links, but the links failed to break apart. A NMFS official acknowledged that weak links are not effective for all types of entanglements. For example, if the whale thrashes around in response to the entanglement and becomes wrapped in the gear, the weak link will not disengage. However, NMFS officials noted that weak links were designed for mouth entanglements, and there have been no documented cases of weak links malfunctioning in a mouth entanglement. Rather, the entanglements with weak links that failed to break apart were entanglements that involved the whale's tail. Even though weak links may not enable whales to free themselves each time they encounter gear, some scientists told us that weak links should be required because they may prevent certain entanglements and are inexpensive and easy for fishermen to use. In fact, two of the three fishing industry association groups with whom we spoke support the use of weak links. The third group, while supportive of using weak links, wanted the breaking strength of weak links to be maintained at its current level during the fall and winter months because if the breaking strength was any weaker, rough tides and weather in offshore waters may cause the buoy to break from the vertical line at the weak link.

Despite their general support of weak links, some of the scientists and conservationists with whom we spoke or who provided written comments to NMFS on the DEIS remain concerned that the breaking strengths of weak links established by NMFS were based on fishing industry needs and not whale protection. According to NMFS scientists, the tests the agency conducted to determine the appropriate breaking strength of weak links were designed to ensure the line does not break when fishermen haul their gear. NMFS officials stated that the agency also considered what was needed to protect whales when developing the breaking strength for weak links. However, research by a scientist at the Stellwagen Bank National Marine Sanctuary and members of the fishing industry suggests that gillnet fishermen could operate successfully using weak links that would be easier for whales to break, specifically a 600-pound breaking strength rather than the current 1,100-pound strength.[33] NMFS officials stated that despite what the report said, the lower breaking strength may not be operationally feasible because after the report was released a fisherman involved in the study experienced failures on some of the weak links in his gear. NMFS officials also questioned whether larger gillnet vessels in deeper water would be able to successfully operate with 600-pound weak links.

Similarly, some of the scientists and conservationists with whom we spoke or who provided written comments to NMFS on the DEIS expressed concern about the areas NMFS proposed for exemption from the gear modifications.[34] Some cautioned that there are risks associated with any exemption area because it only takes one whale traveling within exempted waters for a fatal entanglement to occur—and for right whales one death limits the ability of the species to recover. In addition, some scientists told us that they were concerned that the sightings data used to draw the exemption line may not reflect the actual long-term distribution of whales, as there have been limited efforts to survey the whale population outside of known calving and feeding grounds. In addition, some conservationists note that there have been whale sightings within the exempted areas. However, NMFS officials stated that the agency conducts broad-scale aerial surveys of whales from the Maine-Canada border to New York and has aerial survey coverage in other areas along the east coast as well. In addition, NMFS said in the DEIS that it plans to monitor whale sightings in exempted areas and assess if gear modifications are necessary in these areas.

NMFS Has Not Resolved Potential Implementation Challenges with Using Modified Fishing Gear in Rocky Ocean Bottom Areas

A controversial aspect of the proposed changes to the ALWTR plan that has yet to be resolved is whether sinking groundline is operationally feasible in rocky ocean bottom areas. NMFS told us that it is operationally feasible to use sinking groundline in all areas, but that fishermen may have to modify their fishing practices. For example, the Massachusetts Lobstermen's Association stated that while fishermen have experienced problems operating in rocky bottom areas off the coast of Massachusetts, they have been able to adapt to using sinking groundline. In contrast, officials from the Maine Lobstermen's Association (MLA) told us that fishermen who operate in rocky ocean bottom areas will not be able to use sinking groundline because it will abrade on the rocky bottom—requiring them to replace their rope too frequently and causing gear loss—and may create safety hazards for fishermen.

To assess the feasibility of using sinking groundline, NMFS gear specialists distributed it to 55 fishermen in Northeast states, including Maine, in 2000.[35] NMFS then formally surveyed these fishermen to assess the performance of the sinking groundline in 2003.[36] The 25 fishermen who responded to the survey reported mixed views on the performance of the sinking groundline, with the greatest amount of negative feedback coming from fishermen who operate in eastern Maine. Fishing industry representatives told us that the waters off the coast of eastern Maine consist of rocky bottom. Some of the fishermen who responded to the survey reported experiencing rope abrasion when using sinking groundline in rocky ocean bottom areas. NMFS gear specialists stated that there was a wide range in the length of time that fishermen used the line that was distributed to them in 2000—while some stopped using it after 1 week due to abrasion, others are still using the line today, including some in the rocky bottom areas of Maine. The agency maintains that while fishermen will experience different rates of abrasion in different areas, overall, abrasion will not be a significant problem because fishermen move around and operate in multiple bottom types, instead of exclusively fishing in one area. In addition, NMFS officials noted that rope abrasion is not a problem exclusively associated with the use of sinking groundline; fishermen who use floating groundline also experience rope abrasion.

In addition, NMFS gear specialists maintain that fishermen will be able to use sinking groundline once they gain experience using it. NMFS gear specialists attributed the increased negative feedback regarding using sinking groundline in the rocky areas of Maine to the fact that fishermen in these areas are less likely than fishermen elsewhere to have experience using sinking groundline. The gear specialists told us that fishermen may have to modify their fishing practices in order to successfully use sinking groundline, although NMFS did not discuss this in the DEIS. For example, when using sinking groundline, fishermen will have to be more precise when positioning their boat to haul up their traps. According to these gear specialists, one technique that fishermen could use is to set their boats directly above the traps, so that the fishermen can haul the line straight up and prevent it from getting caught on rocks. However, NMFS maintains that there is no one answer to successfully fishing with sinking groundline on rocky bottom, and it will take fishermen several attempts and techniques to adjust to using sinking groundline.

In contrast, the MLA conducted some limited testing of experimental sinking groundline[37] under contract with the Consortium for Wildlife Bycatch Reduction[38] and concluded that it was not feasible to use in all areas. According to an MLA official, some Maine fishermen reported that sinking groundline performed well, but fishermen who fish in rocky areas generally reported negative experiences. An MLA official told us that, due to abrasion, sinking groundline does not last longer than 1 month in the rockiest areas of Maine, where fishermen experienced such bad abrasion that they stopped using the line to avoid losing their traps. At best, in areas of Maine that are not as rocky, the MLA official told us that sinking groundline would last 1 year— 5 years less than NMFS's estimate in the DEIS.[39] However, the MLA acknowledged that sinking groundline was only tested for a short period of time and therefore recommends additional testing to get a better understanding of its durability.

Fishermen are also concerned that rope abrasion from using sinking groundline in rocky bottom areas will cause gear loss. Based on his professional experience, an MLA official told us that Maine fishermen who fish in rocky bottom areas will experience more gear loss than NMFS estimated because the weakened rope will cause the traps to easily separate. NMFS recognizes that gear loss will be higher, in certain areas, if sinking groundline is required, but a NMFS official told us that rope abrasion will not cause more gear loss than fishermen currently experience because fishermen have the ability to recognize when their rope should be replaced. The NMFS official maintained that fishermen need to be vigilant about the condition of their rope—whether it is floating groundline or sinking groundline—and replace it, as needed, in order to avoid gear loss. However, the agency recognizes that sinking groundline could contribute to increased gear loss as a result of line wrapping around rocks or other marine debris on the ocean floor. If the line becomes caught on the ocean floor, it may break as it hauled to the surface, causing the traps to become separated from the vertical line. When traps become separated from the vertical line, NMFS officials told us that it may be more difficult for fishermen to retrieve their gear if they are using sinking groundline. For Maine inshore fishermen, lost traps will also be more difficult to retrieve because (1) these fishermen are more likely to use shorter trawls than fishermen in other areas—which can be more challenging to locate than a longer trawl that covers more area—and (2) the hook used to retrieve lost gear can bounce off of the rocky bottom, instead of grasping the gear. While an MLA official did not dispute that the factors NMFS cited will contribute to gear loss, he maintained that rope abrasion will also cause gear loss.

MLA officials told us that the Association also has concerns about hauling gear in the manner NMFS described and indicated that there are safety issues with using sinking groundline in rocky bottom areas. Due to rough tidal and weather conditions, an MLA official told us that it is not possible for fishermen to haul their traps from a precise location, as NMFS described. The MLA also is concerned that using sinking groundline in the rocky bottom areas of Maine poses safety issues. For example, if fishermen attempt to haul line that is caught on a rock, their boat could tip, potentially causing injury. Also, if the line snaps when being hauled because it has been weakened due to abrasion, it could strike and injure people on the boat. The Atlantic Offshore Lobstermen's Association also expressed concern about the safety hazards associated with hauling traps using an abraded line that may break. In the DEIS, NMFS acknowledged that there are potential safety hazards associated with the use of sinking groundline. However, an agency official told us that floating groundline can also pose a similar type of safety hazard.

To overcome the operational difficulties associated with using sinking groundline in rocky bottom areas, the Maine Department of Marine Resources submitted a proposal to NMFS in January 2007 that outlined an alternative to the use of sinking groundline along rocky areas of Maine's coast. One of the most prominent features of this proposal involves using low-profile groundline instead of sinking groundline in Maine's rocky bottom areas. Low-profile groundline is still in development, but to reduce abrasion, the Department of Marine Resources tested a line that floats, on average, about 3 feet above the ocean bottom instead of sinking to the bottom. Maine officials acknowledge that whales are present in the waters where they proposed using low-profile line, but maintain that it is a better alternative to using sinking groundline in rocky bottom areas. The state believes that low-profile groundline will be beneficial for fishermen in these areas, while also protecting whales from entanglement. The scientists with whom we spoke were not willing to support low-profile groundline until further research is conducted because they were unsure if it would reduce the risk of entanglement. NMFS is also concerned because although the low-profile groundline tested by the Maine Department of Marine Resources may on average float 3 feet above the ocean floor, in reality the rope moves constantly in the water, sometimes higher than 3 feet and sometimes lower. When it moves above the average height it could form an arc similar to that of floating groundline creating an entanglement risk for large whales. A NMFS official told us that the agency plans to compile proposals on issues related to overcoming the operational difficulties associated with using sinking groundline, including the Maine Department of Marine Resources' low-profile groundline proposal, and will circulate them to the ALWTR Team for comment and discussion.

NMFS Did Not Adequately Represent Uncertainties Associated With Proposed Gear Modifications Cost and Could Not Fully Assess Impacts on Potentially Affected Fishing Communities

NMFS did not have verifiable data for some of the key variables used in its assessment of the fishing industry's costs of complying with the proposed gear modifications.[40] In lieu of such data, NMFS relied on data that contained significant uncertainties about the compliance costs. NMFS acknowledged these uncertainties but, by not analyzing and presenting a range of possible costs, did not adequately represent them in the cost assessment included in the DEIS. As a result, the extent to which the fishing industry's actual costs to comply with the proposed gear modifications could be lower or higher than the amount reported in the DEIS is unclear. In addition, NMFS could not fully assess the impacts of these costs on fishing communities because it lacked data to estimate the affected fishermen's ability to absorb additional compliance costs as well as which specific communities would have to absorb any loss in jobs. Without such data, the agency could not adequately determine how many fishermen would be forced out of business or what impact this would have on communities whose economies are closely tied to the fishing industry.

Significant Uncertainties Exist Regarding NMFS's Cost Estimates of Complying with the Proposed Gear Modifications

NMFS estimated that the total cost to the fishing industry of complying with the proposed gear modifications would be about $14 million annually.[41] NMFS estimated that the lobster industry would incur more than $12.8 million of the projected $14 million costs. To estimate these costs, NMFS analyzed important differences between fishermen such as their location of operation, number of months of operation, and what they catch. This approach allowed the agency to capture variations in the gear configurations and operating characteristics of different types of fishermen and their associated differences in expected compliance costs. NMFS also identified and analyzed the key variables that are responsible for the total cost of complying with the proposed gear modifications, such as the lifespan of groundline, price of groundline, amount of gear loss, and the number of fishermen that would incur these costs. However, there were significant uncertainties associated with the data used to develop these cost estimates, which were not fully represented in NMFS's single cost estimate.

First, NMFS determined the lifespan of both floating and sinking groundline based on undocumented estimates from fishermen and commercial marine suppliers it interviewed, rather than data that could be verified from field tests of groundline. Knowing the lifespan of groundline is important because replacing it more frequently results in higher costs to fishermen. Though NMFS tested sinking groundline to determine if it was operationally feasible to use throughout the northeast coast, it did not use the results of these tests to determine its lifespan. The agency believes that field testing would not have provided better information than the interviews it conducted on the lifespan of groundline because its use varies from fisherman to fisherman. Based on its interviews, NMFS reported in the DEIS that sinking groundline, depending on its diameter, would last between 1 and 3 years less—a 17 to 33 percent shorter lifespan—than the corresponding diameter of floating groundline.[42] However, NMFS could not provide documentation of its interviews or details on how the lifespan—as reported by those interviewed—varied. According to the MLA, the lifespan of sinking groundline can range substantially and could be much shorter than the average NMFS reported in the DEIS. In the DEIS, NMFS acknowledged that the lifespan of groundline is extremely uncertain due to variations in where it is used, such as water temperature and bottom conditions, and the specific operating practices of fishermen. NMFS does not expect that all fishermen's groundline would have the same lifespan as the estimates reported in the DEIS and acknowledges that actual costs to replace groundline could be higher or lower than estimated. Nonetheless, the agency believes that its estimates of the lifespan of sinking groundline are accurate and reflect what fishermen would experience in typical operating conditions. However, by using an average lifespan of groundline in its cost estimate, rather than the range of data collected from fishermen, NMFS did not fully address the concern that the useful life of groundline can vary significantly, depending on a fisherman's practices and fishing location.

Second, while the price of groundline can vary substantially, NMFS did not use a range of prices in its analysis to account for these differences. In 2003, NMFS contacted four commercial marine suppliers and dealers to obtain prices of both sinking and floating groundline. The agency used the median reported price to estimate the costs of replacing floating groundline with sinking groundline. However, the agency does not have documentation of the prices collected and could not describe how these prices varied. We contacted the same suppliers and dealers and found that the price of groundline can range substantially. For example, in February and April 2007 the price of 3/8" sinking groundline—the most commonly used groundline by fishermen and within NMFS's cost analysis—ranged from almost 1 percent to almost 34 percent higher than the price reported in the DEIS.[43] NMFS acknowledges that the price of groundline could be higher or lower than reported in the DEIS but did not analyze and report the range of groundline prices it collected from suppliers and dealers.

Third, NMFS's estimates of the costs of gear loss were based on expert opinion because data from field tests were not available. In the DEIS, NMFS generally reported that fishermen that comply with the proposed gear modifications would experience greater gear loss than they do currently. For example, sinking groundline could lead to greater gear loss because the groundline can get caught on rocks and break as gear is hauled up. However, due to a lack of data, NMFS cannot estimate with confidence how much gear loss would increase for fishermen complying with the proposed gear modifications. The agency did not believe it would be practical to conduct field testing to determine what gear loss could be throughout the Atlantic because it can vary greatly, depending on how and where the gear is used. Instead, NMFS relied on the expert opinions of its gear research team, composed of ex-fishermen who are experienced with fishing gear, and the contractor that prepared the DEIS to estimate gear loss. The research team and the contractor assumed that gear loss attributable to the proposed gear modifications would be approximately double what the fishing industry currently loses in most areas. They estimated that gear loss would be even higher—approximately three times as much as they currently lose—for fishermen operating in areas near the coast of Maine due to difficulties with retrieving gear in rocky bottom areas. While NMFS believes its estimates were reasonable, the MLA believes that these gear-loss estimates are inaccurate and likely to be too low in Maine's rocky bottom areas. The agency does acknowledge that actual gear-loss costs could be higher or lower than it estimated in the DEIS. However, by not analyzing and reporting a range of possible gear-loss costs, NMFS did not fully represent the uncertainty of its gear-loss assumptions, even though it recognized that gear loss can vary, depending on the conditions of use.

Fourth, NMFS may have underestimated the number of Maine lobster fishermen that would be required to comply with the proposed gear modifications. While all fishermen that operate in northern federal waters would be subject to gear modification requirements, all fishermen that operate in state waters along the east coast would not share these requirements because NMFS proposed that some areas be exempted from the regulation.[44] However, NMFS lacked data to effectively determine where state-permitted fishermen operate throughout the year and specifically how many would operate in waters exempted from the new requirements because Maine does not require fishermen to report where they operate.[45] Without this information, NMFS assumed that the percentage of fishermen who would operate in areas exempt from the proposed regulation would correspond to the percentage of state waters that are exempt. For example, NMFS reported in the DEIS that approximately 50 percent of Maine's state waters would be exempted from the gear-modification requirements. The agency also assumed that fishermen would operate in the same areas year-round so those operating in exempted waters would not be affected by the proposed gear modifications. NMFS made this assumption because it believes that lobster fishing in Maine is extremely territorial, and therefore the distance that fishermen move their gear is limited by traditional fishing area boundaries. Consequently, the agency assumed that approximately 50 percent of Maine's lobster fishermen, or approximately 1,853 fishermen, would operate exclusively in exempted waters and would not be affected by the gear-modification requirements.[46] However, a Maine state official and a MLA representative told us that it was unreasonable to assume that lobster fishermen would operate in only one area throughout the year. In fact, they said that fishermen operate wherever lobsters are, which may be in or out of exempted waters. If so, NMFS may not have captured the costs of the proposed gear modifications for an unknown number of Maine fishermen, and therefore may have underestimated how many would be affected by the proposed ALWTR plan changes and thus the total associated costs to the fishing industry.

NMFS acknowledges that there were uncertainties with the data used in its analysis of the costs to the fishing industry and that actual costs could be higher or lower than presented in the DEIS. However, NMFS did not determine the extent to which changes in the lifespan of groundline, price of groundline, amount of gear loss, or the number of fishermen who would have to comply with these requirements would impact the overall $14 million cost estimate. By reporting a single estimate rather than a range of the fishermen's compliance costs, the DEIS did not adequately represent the uncertainties of these key variables in NMFS's assessment. Furthermore, without reporting such a range to account for these uncertainties, the extent to which the total estimated cost of complying with the proposed gear modifications could be different than the $14 million estimate reported in the DEIS is unclear.

NMFS Could Not Fully Assess the Impacts of the Proposed Changes on Fishing Communities Because It Lacked Data on Fishermen's Ability to Absorb Additional Costs and Remain in Business

In addition to assessing the cost of the proposed gear modifications to the fishing industry, NMFS analyzed the effects of the costs of complying with the proposed gear modifications on both fishermen and fishing communities. Conducting an analysis of the effects on fishing communities first requires determining fishermen's ability to absorb additional costs and remain in business and may also include an estimate of changes in regional employment and income directly and indirectly related to the cost of complying with the proposed regulation.[47] However, NMFS could not fully conduct these analyses due to a lack of data.

Specifically, NMFS lacked data on fishermen's costs and revenue in a way that it could estimate their ability to absorb the increased costs of complying with the proposed gear modifications without going out of business. Instead, NMFS estimated fishermen's average annual revenue and then made an arbitrary assumption about the level of increased costs that would cause a fisherman to go out of business. First, NMFS estimated fishermen's annual revenue based on a limited number of fishermen because comprehensive revenue data do not exist. For example, NMFS used data from 9 lobster fishermen to estimate the revenue of 284 northern lobster fishermen that operate vessels less than 28 feet long. However, without fishermen-specific revenue data for all fishermen, the agency was unsure how well its estimates would compare with their actual revenue. Regarding small lobster vessels, NMFS said that it is possible that its analysis in the DEIS systematically underestimates their revenue. NMFS then made an arbitrary assumption that if gear-modification costs were greater than 15 percent of a fisherman's estimated annual revenue, then the fisherman could not absorb the additional costs and would go out of business. NMFS reported in the DEIS that it made this assumption because there is no clearly defined threshold of additional costs that would cause a fisherman to go out of business. Using this assumption, NMFS estimated that approximately 379 fishermen would go out of business, including many that operate smaller vessels for which NMFS lacked actual revenue data. However, because fishermen's actual revenues, as well as their ability to absorb additional costs, could be noticeably different from what NMFS assumed, the number of fisherman that would go out of business could be lower or higher than NMFS estimated.

Furthermore, because NMFS lacked information about which specific fishermen, living in which communities, would go out of business, it could not predict the extent to which specific communities would be affected. That is, NMFS could not identify exactly which communities would lose jobs or quantify any loss of regional income as the result of complying with the regulation. NMFS officials stated that associating any impact to a particular fishing community is particularly difficult because fishermen can sell their fish in one town, harbor their boat in a different town, and reside in a neighboring town. As an alternative, the agency identified potentially affected counties that had (1) over 100 fishermen that would be subject to the ALWTR plan requirements and (2) reported annual landings—seafood caught by fishermen—over 2 million pounds by vessels using ALWTR plan regulated gear.[48] The agency identified 15 counties that met these criteria, many of which were in Maine and economically dependent on the fishing industry.[49]

The agency reported a general description of possible employment effects on these counties, but could not quantify and specifically associate the impact of lost income and employment to any specific community. Consequently, it is not clear how significant the potential economic impacts on these communities would be and how well these communities could withstand the potential loss of fishing jobs and related income.

NMFS Has Not Developed Strategies for Fully Evaluating the Effectiveness of the Proposed Gear Modifications

Although NMFS's proposed modifications to the ALWTR plan contain some revisions to the current gear-marking requirements, such as increased marking of the vertical line, the agency has not developed a comprehensive approach to gear marking that would provide more complete information about the gear involved in future whale entanglements. Markings on commercial fishing gear can enable researchers to assess the type of rope involved in an entanglement, thereby providing critical information to assess the effectiveness of current whale protection measures and insights into needed changes. In addition, NMFS has not developed a strategy for determining whether future entanglements are due to industry noncompliance with the gear modification requirements or the ineffectiveness of the gear modifications themselves.

Lack of Comprehensive Gear-Marking Requirements Could Hamper Assessment of Proposed Gear Modifications

Research on the nature and source of whale entanglements is challenging in that entanglements are not directly observed when they occur. Instead, NMFS's gear research team is forced to rely on the gear it retrieves from entangled whales and/or photographs taken of the entanglement, if any. Even when gear is recovered, the gear research team may have only a rope fragment to evaluate. Therefore, markings on gear can play a critical role in informing scientists about the nature of the entanglement. Gear markings can potentially indicate whether a whale became entangled in groundline or vertical line, whether the gear was from the lobster fishery or some other fishery, and the geographic area where the gear was set. Currently, gear markings, such as vessel or permit numbers on buoys, can identify the name of the fisherman who set the gear so that NMFS officials can obtain specific information from the fisherman, such as the exact location where the gear was set.

Under the current regulation, NMFS requires some trap and gillnet fishermen to place one color-coded, 4-inch mark on the vertical line mid-way through its length, which fishermen typically paint on or tape to the rope. The color-coding scheme provides information about the location and fishery involved in the entanglement. For example, lobster fishermen in the Cape Cod Bay Restricted Area in federal waters—an area NMFS has designated as a critical habitat for large whales—are required to use a red mark. Other colors are used to indicate other fisheries and areas. However, according to a NMFS official, the current gear-marking scheme has not been effective in assisting researchers because only rarely have fragments of vertical line been recovered that included the required mark.

NMFS proposed a new requirement for marking vertical line because the agency recognized that markings would be useful as the agency and the ALWTR Team further evaluate vertical line for future regulatory action. For example, if the agency recovered a rope that was marked, it would be better able to determine that it was vertical line and how frequently vertical line was involved with entanglements. Specifically, NMFS proposed expanding the frequency of gear marking—to one 4-inch mark every 60 feet on the vertical line.[50] A NMFS official with whom we spoke said the agency based the 60-foot requirement on the average length of rope found on entangled whales. The official explained that the 60-foot requirement would increase the likelihood of recovering marked rope from an entangled whale and would also minimize the burden on fishermen by not requiring them to mark rope even more frequently.

However, we believe NMFS's proposed gear-marking requirement may not be adequate in assisting researchers in identifying the gear that is recovered from an entangled whale because it is not comprehensive. First, even with increased markings on vertical lines, researchers may still not retrieve the marked portion of the rope. For example, some of the rope recovered from entangled whales has been only 6 feet long. Some stakeholders, including scientists at the Provincetown Center for Coastal Studies, recommended that NMFS require continuous marking throughout the length of the rope through the use of tracer lines—colored threads of line throughout the length of the rope. However, according to a NMFS official, continuous marking throughout the length of the rope is not practical because, among other reasons, it would limit fishermen's ability to move between different fishing areas that require different color markings.

Second, NMFS has not proposed marking sinking groundline. NMFS did not provide a rationale in the DEIS for not requiring the marking of sinking groundline. However, a NMFS official told us that the agency believed that sinking groundline would be completely effective at reducing groundline entanglements, and therefore there was no need to burden fishermen with a marking requirement on such line. However, scientists with whom we spoke believe that while sinking groundline will reduce entanglement risk, they do not believe that its complete success can be guaranteed. For example, scientists have observed endangered whales with mud on their heads, which scientists believe whales acquired scraping the ocean floor as they feed. Based on this information, scientists are concerned that endangered whales could become entangled in sinking groundline. Consequently, several scientists with whom we spoke, including several NMFS scientists, told us that sinking groundline should be marked so its performance can be evaluated.

Third, NMFS did not require gear markings in areas that have been exempted from the proposed gear modifications. NMFS developed exempted areas because the agency determined, based on whale sighting data, that certain waters pose a relatively low risk of entanglement because they are not as frequently traveled by endangered whales as others. However, because some of these areas are dense with commercial fishing gear, they nevertheless pose some risk. Consequently, we believe that any assessment of the new regulations would benefit from gear markings on the gear used by fishermen in exempted areas, even if they are not required to use modified gear.

Various stakeholders with whom we spoke or who submitted comments on the DEIS expressed concern about NMFS's proposed gear-marking scheme. Industry representatives were concerned about the burden the requirement would place on fishermen who would have to mark rope more frequently and the impracticality of marking rope every 60 feet. According to the Massachusetts Lobstermen's Association, painted marks can fade or become covered by algae and therefore must be maintained to retain their visibility—a problem that would be exacerbated with additional marking requirements. Also, maintaining a 60-foot space between marks is difficult because commercial fishermen must routinely cut and splice fishing lines. For example, fishermen may find their ropes inadvertently cut due to commercial and recreational vessel traffic and need to splice rope together. Fishermen may also change the length of their ropes when moving gear into and out of deep water. Given the impracticality of marking rope every 60 feet, the Cetacean

Society International stated that NMFS should consider requiring rope that was marked continuously through the length of the rope by the manufacturer.

Stakeholders with whom we spoke observed that markings that were specific to individual fishermen could be useful to researchers because they would enable researchers to obtain information from fishermen, specifically on how and where they set their gear. The Maine Lobstermen's Association and the Provincetown Center for Coastal Studies noted that new technology, such as microchips embedded in fishing rope, could potentially provide fishermen-specific information and that they would favor its use if the technology was feasible in the commercial fishing environment. NMFS's gear research team is aware of this technology, but believes that it is not yet suited to commercial fishing conditions because microchips embedded in rope may pop out as the rope moves through hauling devices used to pull gear out of the water.

NMFS Lacks a Strategy for Assessing Industry Compliance with the Proposed Gear Modifications

NMFS has not developed a strategy that will allow it to determine whether any future whale entanglements are due to noncompliance with the proposed new gear requirements by fishermen or the ineffectiveness of the gear modifications. NMFS did not specify in the DEIS how it plans to monitor industry compliance with its proposed rule and has not developed such a plan outside of the DEIS. Stakeholders with whom we spoke or who submitted comments on the DEIS have expressed concern that the DEIS did not include a plan for monitoring compliance with the proposed rule. According to the Whale Center of New England, the lack of monitoring plans in the past have made it difficult to evaluate the effectiveness of previous actions, and as a new regulation goes into effect, a monitoring plan would be critical in assessing the success or failure of the proposed actions.[51] A Provincetown Center for Coastal Studies scientist observed that a plan for monitoring the proposed rule is as important to effectively protecting whales as the gear modifications themselves. A NMFS official told us that the agency understands the importance of having a compliance strategy and plans to develop one.

Regarding the current regulatory requirements, NMFS has not conducted a systematic survey of industry compliance and therefore, does not know the extent of industry compliance along the east coast. Maine is the only state to have conducted even limited compliance surveys of its state-permitted vessels. Since 2002, Maine has conducted annual compliance surveys over a 30-day period in both state and federal waters off the coast of Maine, according to a Maine Department of Marine Resources official.[52] During the survey, enforcement officers in patrol boats target large concentrations of gear and randomly pull gear out of the water. The enforcement officers document information about the type and location of gear, the owner, and what species the fishermen were targeting. This effort is conducted separately from routine enforcement patrols during which enforcement officers complete logs that record only violations. According to a Department of Marine Resources official, the state can conduct this compliance survey because it has vessels that are equipped to haul commercial fishing gear, even from deep water areas and because NMFS has provided funding to support this effort. Although Maine's annual compliance survey indicates a high rate of compliance, it is subject to a number of limitations. The survey is not conducted using scientific sampling of gear, so the results cannot be generalized to all gear, and it does not incorporate all segments of Maine's fishing industry, so it is not comprehensive.

Effective January 2007, Massachusetts required that sinking groundline be used throughout state waters—a requirement similar to what NMFS proposed along the north Atlantic coast. Officials with the Massachusetts Office of Law Enforcement Environmental Police stated that they are exploring the use of a vessel equipped with sonar to assess whether fishermen are complying with the state's sinking groundline requirement. Through sonar, the department can detect if fishermen are using floating or sinking groundline without hauling the gear out of the water. They explained that sonar could be an efficient method for conducting a systematic survey because hauling gear is time consuming, particularly since the gear must be placed back carefully where the fisherman had the gear set.

CONCLUSION

NMFS has a challenging mandate of reducing the risks posed to the survival of Atlantic large whales by entanglements in commercial fishing gear, while also taking into account the economic interests of commercial fishermen. In its continuing efforts to protect endangered whales, including the right whale which is critically endangered, NMFS is considering various revisions to the existing regulations which include certain gear modifications for the fishing industry. However, the economic analysis that NMFS developed to support its actions does not disclose the full range of possible costs that the proposed gear modifications may impose on fishermen although it acknowledges that costs could be higher or lower than it presented. While we believe the approach that NMFS used to estimate compliance costs is reasonable, we are concerned that the presentation of costs did not fully reflect the uncertainty of the analysis. Moreover, given the concerns raised by scientists and other experts regarding the effectiveness of the proposed gear modifications in eliminating whale entanglements, it is important for NMFS to develop strategies that will allow it to assess the effectiveness of these changes as well as monitor industry compliance. However, NMFS has neither developed a comprehensive strategy to help it assess whether its proposed gear modifications are effective in eliminating whale entanglements nor has it developed a program to monitor industry compliance.

RECOMMENDATIONS

Before NMFS finalizes its proposed regulations for the ALWTR plan, we recommend that the Secretary of Commerce direct the Administrator of National Oceanic and Atmospheric Administration to direct the Assistant Administrator for NMFS to take the following three actions:

- adequately represent the uncertainty in data that the agency used to determine the costs of the proposed fishing gear modifications, by presenting a range of possible costs in the economic analysis section of the final environmental impact statement;
- revise the proposed gear-marking requirements to include markings on sinking groundline and gear marking requirements in exempted areas; and
- develop a strategy for assessing the extent of industry compliance with the gear modification requirements.

AGENCY COMMENTS AND OUR EVALUATION

We provided a draft of this chapter to the Department of Commerce for review and comment. In its comments, the Department of Commerce's NOAA questioned whether we had obtained input from a broad range of stakeholders, felt the report appeared to focus solely on the impacts to the Maine fishing community, and disagreed with two of our recommendations.

We disagree with NOAA's comment that we did not obtain and reflect a range of stakeholders' views in this chapter. As described in detail in our objectives, scope, and methodology, included in appendix I, we conducted interviews, reviewed documents, and took other steps to ensure that our work adequately portrays a wide range of stakeholders' views and appropriately addresses the complexities of these issues. In addition to NMFS officials, the stakeholders we contacted include state marine fishery management agency officials from Maine and Massachusetts; fishing industry representatives from the Massachusetts Lobstermen's Association, Maine Lobstermen's Association, and the Atlantic Offshore Lobstermen's Association; a representative from the Humane Society of the United States; and scientists from the Provincetown Center for Coastal Studies, the New England Aquarium, the Woods Hole Oceanographic Institution, and the Marine Mammal Commission. We also reviewed all of the stakeholders' comments submitted to NMFS on the DEIS and attended a meeting of the ALWTR Team—composed of fishermen, scientists, conservationists, and state and federal officials who are tasked with monitoring the status of the ALWTR plan and advising NMFS as it develops revisions to the plan.

In its general comments, NOAA also stated that, in its view, the draft report appears to focus solely on the impacts to the Maine fishing community. We do not agree with this characterization of the report. Although the report clearly places some emphasis on issues of concern to the Maine lobster industry, we believe this is appropriate given the objectives we were asked to address in this review. Two of our objectives specifically focus on how NMFS plans to address issues related to implementing the proposed changes to the ALWTR plan, particularly in the rocky bottom areas of the north Atlantic coast, and to evaluate the extent to which NMFS fully assessed costs to the fishing industry and economic impacts on fishermen. The rocky bottom areas of concern are located primarily off the coast of Maine; and as a result, the report describes concerns raised by Maine lobstermen regarding the implementation challenges they believe they will face. In addition, according to NMFS's analysis contained in the DEIS, the lobster industry will bear $12.8 million of the approximately $14 million annual cost of complying with the proposed regulatory changes, and this industry is primarily centered in Maine and Massachusetts. Consequently, the report appropriately includes concerns raised by Maine lobstermen about NMFS's analysis of the costs of complying with the proposed regulatory changes.

With regard to our recommendations, NOAA believes that the uncertainty of the data was adequately represented in the DEIS and therefore did not agree with our recommendation that the agency present a range of possible costs in its final economic analysis to represent the uncertainty in the data. Nonetheless, NOAA said that it is planning to clarify the variations and uncertainties within its analysis in the Final Environmental Impact Statement (FEIS). NOAA said that this clarification would discuss potential differences in total compliance cost from variations in several of the assumptions that we had identified in our report. By recognizing that the treatment of uncertainty in the DEIS can be improved and by taking additional steps to explain the effect of uncertainty on compliance costs, the agency appears to be taking a step in the direction we recommended. However, we continue to believe that unless NMFS includes a range of possible costs facing the fishing industry in the FEIS, the agency will not have clearly and thoroughly represented the uncertainties in its analysis.

NOAA also did not agree with our recommendation that the agency revise the proposed gear-marking requirements to include markings on sinking groundline and gear marking in exempted areas. Although NOAA concurred that methods are needed for identifying sinking groundline and gear from exempted areas, it stated that such markings are not feasible or practical at this time. It is unclear to us why NOAA would make such a statement given that in the DEIS, NMFS has proposed marking requirements for vertical line. We believe that if such marking is feasible and practical for vertical line, similar marking should also be feasible and practical for sinking groundline. Without such comprehensive gear- marking requirements, we believe that NMFS will not be in a position to evaluate whether or not its regulations, including the use of sinking groundline, will be effective in protecting Atlantic large whales from entanglement.

NOAA did agree with our recommendation that NMFS should develop a strategy for assessing industry compliance with the gear-modification requirements. However, NOAA stated that the recommendation cannot be implemented before NMFS finalizes its proposed regulations for the ALWTR plan, as we recommended. This is because NMFS is currently working on the strategy and plans to continue discussions with the ALWTR team at its next meeting, tentatively scheduled for early 2008, which is beyond the time the FEIS and final regulation will be issued. If NOAA is unable to complete its strategy for assessing industry compliance prior to finalizing its proposed regulations, we believe the agency should have the strategy in place by the effective date of the final regulations so that NMFS can be in a position to evaluate the effectiveness of its regulatory changes from their inception.

NOAA also provided technical comments, which we have incorporated in this chapter as appropriate. NOAA's comments and our detailed responses are presented in appendix II.

Sincerely yours,

Anu K. Mittal

Anu K. Mittal
Director, Natural Resources and Environment

APPENDIX I. OBJECTIVES, SCOPE, AND METHODOLOGY

Since the National Marine Fisheries Service (NMFS) has not issued a final environmental impact statement or regulation, we have reviewed the proposed changes to the Atlantic Large Whale Take Reduction (ALWTR) plan outlined in the draft environmental impact statement (DEIS). Specifically we (1) described the scientific basis for the proposed changes to the ALWTR plan and the extent to which uncertainties exist regarding how effectively they will protect large whales; (2) described how the agency plans to address implementation issues, particularly in the rocky bottom areas of the North Atlantic coast; (3) evaluated the extent to which NMFS fully assessed costs to the fishing industry and economic impacts on fishermen; and (4) evaluated the extent to which NMFS has developed strategies for fully assessing the effectiveness of and industry compliance with the proposed changes.

To address all four objectives, we reviewed the DEIS and the public comments made in response to the issuance of the DEIS. We interviewed officials at NMFS's Northeast Regional Office who participated in developing the proposed changes to the plan outlined in the DEIS. We interviewed state marine fishery management agency officials from the Maine Department of Marine Resources and the Massachusetts Division of Marine Fisheries. We also interviewed fishing industry representatives from the Massachusetts Lobstermen's Association, Maine Lobstermen's Association, and the Atlantic Offshore Lobstermen's Association. We interviewed a representative from the Humane Society of the United States, a conservation group. Finally, we attended a meeting of the ALWTR Team—composed of fishermen, scientists, conservationists, and state and federal officials who are tasked with monitoring the status of the ALWTR plan and advising NMFS as it develops revisions to the plan—held in December 2006. At this meeting, we observed presentations on the status of endangered whales, new strategies to reduce the entanglement risk of vertical line, and experimental whale safe rope that could be used in rocky bottom areas.

To obtain information on the scientific basis for the proposed changes to the ALWTR plan outlined in the DEIS and any uncertainties regarding how effectively they will protect large whales, we interviewed NMFS scientists at its Northeast Fisheries Science Center and officials that developed the proposed changes to the ALWTR plan. In addition, we interviewed marine mammal scientists from the Provincetown Center for Coastal Studies, the New England Aquarium, the Woods Hole Oceanographic Institution, and the Marine Mammal Commission to obtain expert opinions on the need for regulatory action and the effectiveness of the actions proposed by NMFS. We also reviewed scientific literature on right, humpback, and fin whale behaviors and entanglements in commercial fishing gear. Additionally, we attended the annual meeting of the North Atlantic Right Whale Consortium, a group composed of a number of both nongovernmental and governmental organizations and individuals, including marine mammal scientists, who study and work to conserve North Atlantic right whales. At this meeting, we observed presentations on recent scientific research on right whales, including their migratory behaviors and entanglement risks.

To obtain information on how NMFS plans to address issues with implementing its proposed changes to the ALWTR plan, especially in rocky bottom areas of the North Atlantic coast, we obtained the opinions of representatives from fishing industry associations on the challenges posed by the proposed gear modifications. We also interviewed NMFS officials from its gear research team—former fishermen who develop whale safe gear and provide educational outreach to fishermen—to obtain information on how fishermen could overcome these challenges. In addition, we interviewed officials from the Maine Department of Marine Resources and the Massachusetts Division of Marine Fisheries to obtain their views on how these challenges could impact fishermen. Finally, we reviewed the results from NMFS's testing of sinking groundline throughout the east coast as well as the results of the Maine Lobstermen's Association's tests of sinking groundline.

To evaluate the extent to which NMFS fully assessed the costs to the fishing industry and impacts to fishing communities, we interviewed representatives of Industrial Economics Inc., the contractor who conducted the economic analysis that is included in the DEIS. We also interviewed officials from NMFS's Northeast Regional Office, including the gear research team, that contributed to the economic assessment. In addition, we interviewed economists from NMFS's Northeast Fisheries Science Center. We also interviewed fishing industry representatives to get their views on the data and assumptions used within the DEIS analysis. We also contacted commercial marine suppliers in February and April of 2007 to obtain the price of sinking groundline.

To evaluate the extent to which NMFS has developed strategies for assessing the effectiveness of and industry compliance with the proposed ALWTR plan changes, we interviewed officials from NMFS's Northeast Regional Office on NMFS's current and planned evaluation strategies. We interviewed NMFS's gear research team and officials from the Northeast Regional Office that developed the gear-marking scheme on the proposed gear-marking requirements and how they were developed. We interviewed scientists from the Provincetown Center for Coastal Studies, the New England Aquarium, and the Woods Hole Oceanographic Institution to obtain their views on the proposed gear-marking requirements and which markings would be most beneficial to assessing the effectiveness of gear modifications. We also interviewed representatives from the Maine Lobstermen's Association to obtain their views on gear-marking requirements. Finally, we interviewed marine fisheries law enforcement officials from the Massachusetts Executive Office of Environmental Affairs and the Maine Department of Marine Resources on gear-marking requirements and their current compliance evaluation strategies.

We conducted our review from August 2006 through June 2007 in accordance with generally accepted government auditing standards.

APPENDIX II. COMMENTS FROM THE DEPARTMENT OF COMMERCE

Note: GAO comments supplementing those in the report text appear at the end of this appendix.

THE DEPUTY SECRETARY OF COMMERCE
Washington, D.C. 20230

July 5, 2007

Ms. Anu K. Mittal
Director, Natural Resources
  and Environment
U.S. Government Accountability Office
441 G Street, NW
Washington, D.C. 20548

Dear Ms. Mittal:

Thank you for the opportunity to review and comment on the Government Accountability Office's draft report entitled *National Marine Fisheries Service: Improved Economic Analysis and Evaluation Strategies Needed for Proposed Changes to Atlantic Large Whale Protection Plan* (GAO-07-881). On behalf of the Department of Commerce, I enclose the National Oceanic and Atmospheric Administration's programmatic comments to the draft report.

Sincerely,

David A. Sampson

Enclosure

See comment 1.

Department of Commerce
National Oceanic and Atmospheric Administration
Comments on the Draft GAO Report Entitled
"National Marine Fisheries Service: Improved Economic Analysis and Evaluation
Strategies Needed for Proposed Changes to Atlantic Large Whale Protection Plan"
(GAO-07-881/June 2007)

General Comments

The Department of Commerce's National Oceanic and Atmospheric Administration (NOAA) acknowledges GAO's efforts in addressing previous comments provided on the Statement of Facts and recognizes GAO staff for its hard work toward understanding the issues and complexities of managing these high profile protected species.

NOAA has three general comments on the report's content.

1. In considering issues surrounding the conservation of Atlantic large whales, NOAA takes into account a variety of input from diverse stakeholders. In developing the proposed alternatives, for example, NOAA weighed input from the Atlantic Large Whale Take Reduction Team (ALWTRT), which includes:
   - seventeen individuals representing fishing organizations and groups that utilize trap/pot gear;
   - twelve individuals representing fishing organizations and groups that utilize gillnet gear;
   - five individuals representing conservation/environmental groups;
   - thirteen individuals representing state agencies;
   - eight Federal Government and fishery management organizations; and
   - eight academic/scientific organizations.

   In addition, NOAA has discussed many of the management concepts contained in the Draft Environmental Impact Statement (DEIS) at several public forums including:
   - six scoping meetings held prior to the development of the DEIS;
   - several ALWTRT meetings; and
   - thirteen public hearings in which extensive public testimony was provided.

   NOAA also received 81 letters providing comments on the DEIS and received approximately 25,000 additional form letters via e-mail and regular mail.

   NOAA balanced the input received with all of the various legal mandates to which it is required to adhere. NOAA encourages the GAO to ensure the report adequately portrays the various stakeholder views and the complexities involved in addressing these complicated issues.

2. The Atlantic Large Whale Take Reduction Plan (ALWTRP) proposes changes that affect commercial fishing operations from Maine to Florida; however, the GAO draft report appears to focus solely on the impacts to the Maine fishing community, which is only one

See comment 3.
See comment 4.
See comment 5.
See comment 6.
See comment 7.
See comment 8.

sector of the stakeholders affected by this rulemaking. If the intent of the report is to focus on primarily one sector affected by the ALWTRP, the report and its title should be revised to clarify this. However, if the intent is to provide a more comprehensive assessment, other stakeholder views should be included. In addition to the Maine Lobstermen's Association (MLA), whose views seem to have been heavily relied upon by GAO in developing its findings and conclusions, there are other industry-based organizations within the state of Maine. The Downeast Lobstermen's Association, Southern Maine Lobstermen's Association, and Maine Offshore Lobstermen's Alliance are also prominent industry-based organizations within the state of Maine. In addition, the MLA and the other industry-based groups in Maine do not always share the same viewpoint. NOAA notes GAO should have more equally reflected the wide extent of the plan, including geographic extent, range of fisheries affected, conservation interests and other aspects of the plan outside of select stakeholders from Maine. For example, the report would have benefited from input from other industry-based organizations such as the Massachusetts Lobstermen's Association, the Atlantic Offshore Lobstermen's Association, the Garden State Seafood Association, and the North Carolina Fisheries Association. In addition, input from conservation interests like the Humane Society of the United States, the Ocean Conservancy and the International Wildlife Coalition, as well as right whale research organizations like the New England Aquarium would have provided a wider viewpoint, allowing perspectives from other stakeholders affected by the ALWTRP.

3. The report provides numerous statements indicating NOAA's National Marine Fisheries Service (NMFS) did not provide data to support some of its important caveats and estimates used in its DEIS analysis. For example, the report states, "Based on its interviews, NMFS reported in the DEIS that sinking groundline, depending on its diameter, would last between one to three years less—a 17 to 33 percent shorter lifespan—than the corresponding diameter of floating groundline. However, NMFS could not provide documentation of its interviews or details on how the lifespan—as reported by those interviewed—varied." To support the author's claim, the report states, "According to MLA, the lifespan of sinking groundline can range substantially and could be much shorter than the average NMFS reported in the DEIS." NOAA believes this is the opinion of the MLA (with no documentation of how the MLA arrived at its conclusion provided in the GAO report). Consistent with general comments 1 and 2 above, NOAA believes that GAO should have included views from other stakeholders, particularly those who, unlike the MLA, are not proposed to be regulated under this action. Examples of such stakeholders would be those in the environmental and science communities.

See comment 9.
See comment 10.
See comment 11.

NOAA Response to GAO Recommendations

The draft GAO report states, "Before NMFS finalizes its proposed regulations for the ALWTR plan, we recommend that the Secretary of Commerce direct the Administrator of NOAA to direct the Assistant Administrator for NMFS to take the following three actions:"

NOAA Response: NOAA does not believe actions to fully address the recommendations can be implemented "before NMFS finalizes its proposed regulations for the ALWTR plan." NOAA has evaluated the recommendations and has provided details below.

Recommendation 1: "Adequately represent the uncertainty in the data the agency used to determine the costs of the proposed fishing gear modifications by presenting a range of possible costs in the economic analysis section of the final environmental impact statement."

NOAA Response: NOAA believes that it has adequately represented the uncertainty in the data the agency used to determine the costs of the proposed fishing gear modifications. Thus, NOAA does not agree with the recommendation to present a range of possible costs in the economic analysis section of the final environmental impact statement. Given the proposed regulations vary depending on fishery, location of fishing activity, time of year, and the variety of fishing habitats and practices, data are not available to assess differences in gear loss, wear, and replacement rates specific to each vessel or to develop probability distributions. The economic analysis contained in the DEIS relies on the best professional judgment to estimate the average rate of gear usage, replacement, and loss under varying conditions for varying fishing locations.

GAO reports MLA anticipated different rates of gear loss and replacement. Similarly, NOAA anticipates and acknowledges in the DEIS that certain vessels will experience higher rates of gear loss and replacement and, as a result, higher compliance costs. NOAA also anticipates other vessels will experience lower rates of gear loss and replacement and, as a result, lower compliance costs. As such, NOAA believes application of higher gear loss and replacement rates to the entire fishing industry would be misleading.

The report also cites discussions with MLA stating that vessels fishing on rocky bottom within Maine state waters will experience higher rates of gear loss, wear, and replacement than assumed in the analysis. It is important to note, however, that a significant portion of Maine's state waters would be exempt from the provision, including areas of rocky bottom. For vessels fishing in the proposed regulated portions of Maine state waters, the analysis also assumes a higher rate of gear loss than in other waters. As a result, NOAA anticipates the uncertainty in gear loss and replacement within Maine state waters is less than suggested by GAO.

NOAA is planning to clarify the variations and uncertainties within its analysis contained in the Final Environmental Impact Statement. This clarification would discuss potential differences in total compliance costs from variations in several of the assumptions identified in the report.

See comment 12.
See comment 13.

Recommendation 2: "Revise the proposed gear marking requirements to include markings on the sinking groundline and gear marking requirements in exempted areas."

NOAA Response: NOAA does not agree with this recommendation. While NOAA agrees a method for identifying sinking groundline and gear in exempted areas is needed, NOAA does not believe revising the proposed gear marking requirements to include markings on the sinking groundline and gear marking requirements in exempted areas would be feasible or practical at this time. NOAA discussed marking sinking groundline and gear in exempted areas during the development phase of the DEIS and proposed rule. Commenters objected to this gear marking scheme for the following reasons:

1. It would be impossible to adhere to in deep water;
2. Tape will not stick to wet rope, nor will paint;
3. Marking techniques lose their visibility within a few weeks in the water due to basic wear and tear and the accumulation of algal growth on the ropes making the marks hard to discern;
4. Gear marking would be difficult to implement as line is spliced or fouled over the course of its useful life;
5. There would be a problem in trying to figure out whether the space between marks is the exact length;
6. It will be tough to mark at sea, especially given temperature, sea state, and safety considerations;
7. The marking scheme is generic and limited marks will not provide much information. For instance, the scheme would only identify a buoy line or groundline, but not a fishery or even a region where the gear was fished (i.e., no unique identifier); so this limits the amount of information that can be tracked and evaluated;
8. It is too time consuming, costly, impractical, and unworkable;
9. Too many areas will not have marking requirements (e.g., exempted areas, recreational gear, Canadian waters);
10. Gear loss would be too much using the new gear marking;
11. It will be a financial burden to fishermen, without much chance for results that are useful;
12. Buoys and traps are already marked under current lobster fishing rules; and
13. It would be hard to enforce given the large number of recreational lobstermen.

GAO should note NOAA has tested alternative gear marking schemes to address the concerns raised by the industry and is currently working on a chip technology that can be inserted into the line and coded with fishermen identification for the entire eastern seaboard. NOAA anticipates this will help to more easily identify gear in the water. NOAA will be discussing this technology with the ALWTRT in the future. However, NOAA believes it would be premature to propose

See comment 14.

Recommendation 3: "Develop a strategy for assessing the extent of industry compliance with the gear modification requirements."

NOAA Response: NOAA agrees with this recommendation. A strategy should be developed for assessing the extent of industry compliance with the gear modification requirements. However, a strategy cannot be developed prior to NMFS finalizing its proposed regulations for the ALWTRP.

NOAA is currently working on developing a monitoring/compliance strategy with the ALWTRT and other stakeholders. NOAA has discussed this strategy with the ALWTRT on several occasions. However, the results of these discussions were not conducive to development of a meaningful strategy. At its April 2003 meeting, the ALWTRT recommended that NOAA establish a Compliance Committee to discuss issues such as evaluating, monitoring, and improving ALWTRP compliance. The plan development includes working through the Atlantic States Marine Fisheries Commission and Joint Enforcement Agreement (JEA) contacts and involves stakeholder groups on the ALWTRT. As noted in the report, NOAA has made some progress regarding this issue, particularly with NOAA and state enforcement offices through the JEA process. However, NOAA acknowledges more work is needed in this area. At its 2004 and 2005 meetings, the ALWTRT also discussed separating monitoring issues from the Compliance Committee and addressing these through a Status Report Subcommittee. The discussion focused on the interpretations of the annual right whale and humpback whale scarification analysis. Specifically, the ALWTRT discussed whether the scarification analysis was the best method for evaluating the ALWTRP. NOAA intends to continue this discussion with the ALWTRT at its

The following are GAO's comments on the Deputy Secretary of Commerce letter dated July 5, 2007.

GAO Comments

1. We believe that the report reflects a wide variety of input from a diverse group of stakeholders. For this reason, we did not revise the report based on this comment. As discussed in appendix I of the report, we obtained input from stakeholders through interviews, a review of relevant documents, and attendance at meetings. We interviewed fishing industry representatives from the Maine Lobstermen's Association (MLA), the Massachusetts Lobstermen's Association, and the Atlantic Offshore Lobstermen's Association. We obtained the views of the Garden State Seafood Association and the Downeast Lobstermen's Association through the written comments they submitted on the DEIS. We also interviewed officials from Maine's and Massachusetts' state marine agencies because 10 of the 15 communities that the DEIS identified as "at-risk" as a result of the projected economic impacts of the ALWTR plan modifications are located in these two states. We also interviewed a representative of

the Humane Society of the United States and marine mammal scientists at the New England Aquarium, Woods Hole Oceanographic Institution, the Provincetown Center for Coastal Studies, and the Marine Mammal Commission. Moreover, we obtained views from scientists at the NMFS Northeast Fisheries Science Center as well as the views of the NMFS gear research team. We attended the annual meeting of the North Atlantic Right Whale Consortium, a group that studies and works to conserve North Atlantic Right Whales. We also attended the December 2006 ALWTR Team meeting, which included representatives from a wide range of groups including trap and gillnet fishing groups, conservation groups, federal and state agencies, and academic/scientific organizations. Finally, we reviewed all of the public comments submitted to NMFS on the DEIS, which included comments from a wide variety of government, scientific, industry, and environmental groups.

2. We do not agree with National Oceanic and Atmospheric Administration's (NOAA) contention that the report appears to focus solely on the impacts to the Maine fishing community. In addressing the first and fourth objectives of the report, we broadly describe the scientific basis for the proposed changes to the ALWTR plan and evaluate the extent to which NMFS has developed strategies for fully assessing the effectiveness of and industry compliance with the proposed changes. Our second objective was to describe how NMFS plans to address issues related to implementing the proposed changes to the ALWTR plan, particularly in the rocky bottom areas of the North Atlantic coast. These rocky bottom areas are located primarily off of the coast of Maine; and as a result, the report describes concerns raised by Maine lobstermen regarding the implementation challenges they believe they will face. Finally, our third objective was to evaluate the extent to which NMFS fully assessed costs to the fishing industry and economic impacts on fishermen. As NMFS stated in the DEIS, the lobster industry is expected to bear more than $12.8 million of the approximately $14 million annual cost of complying with the proposed regulatory changes, and the Atlantic lobster industry is primarily centered in Maine and Massachusetts. Consequently, the report includes concerns raised by Maine lobstermen about NMFS's analysis of the costs of complying with the proposed regulatory changes. For the reasons stated above, we did not revise the report.

3. As stated in comment two, we do not believe that the report focuses on one industry sector affected by the ALWTR plan. Because we believe that the report title is accurate and appropriate we did not revise the report in response to this comment.

4. We did not rely heavily on the views of the MLA in developing our finding and conclusions as NOAA contends. As we stated in comment one, we made use of information from a wide range of stakeholders in developing our findings. Although the report clearly places some emphasis on issues of concern to the Maine lobster industry, we believe this is appropriate given the objectives we were asked to address in the report. As a result, we did not revise the report in response to this comment.

5. We believe that the report adequately describes the geographic extent of the proposed changes to the ALWTR plan and the range of fisheries affected. In addition, we reviewed comments on the DEIS submitted by the Garden State Seafood Association and obtained input from the Massachusetts Lobstermen's Association and the Atlantic Offshore Lobstermen's Association through interviews with their representatives. We have revised the report to include specific comments from the latter two groups.

6. As we noted in comment one, we interviewed a representative from the Humane Society of the United States and scientists from the New England Aquarium. We also reviewed comments on the DEIS submitted by the Ocean Conservancy and the International Wildlife Coalition. Consequently, we did not revise the report in response to this comment.

7. NOAA correctly states that our report identifies instances in which NMFS could not provide documentation for some of the estimates it used in the economic analysis in the DEIS, including how the lifespan of sinking groundline varied based on interviews NMFS conducted. However, NOAA then erroneously claims that we used statements from the MLA to support the fact that the lifespan of sinking groundline varied. We reported NMFS's contention that the lifespan of sinking groundline varied, despite the fact that it could not provide documentation of the interviews it conducted. We also reported the MLA's view that, based on its experience, the lifespan of sinking groundline can range substantially and could be shorter than the average NMFS reported in the DEIS. For these reasons, we did not revise the report in response to this comment.

8. As stated in comment one, we made use of information from a wide range of stakeholders in developing our findings, including those in the science and environmental communities. However, regarding the costs and economic impacts of gear modifications, we relied on the views of the affected fishermen because they have direct experience in this area, whereas scientists and conversation groups generally do not. Consequently, we did not revise the report in response to this comment.

9. We do not agree that NOAA has adequately represented the uncertainty in the data the agency used to determine the costs of the proposed fishing gear modifications. We believe that presenting its estimates as single point values (for example, $14 million) rather than showing the range of possible costs, implies a degree of preciseness that is misleading and not supportable by the limitations and sometimes lack of available data. Moreover, while, on one hand, NOAA claims that it has adequately addressed uncertainty, on the other hand, it goes on to say that it is planning to clarify the variations and uncertainties within its analysis contained in the Final Environmental Impact Statement. This clarification would discuss potential differences in total compliance costs from variations in several of the assumptions identified in our report. We believe such clarification is needed and continue to believe that presenting a range of possible costs would be the best way to represent the uncertainty in the analysis. For these reasons, we did not revise the report in response to this comment.

10. We agree that gear loss and replacement and the associated compliance costs could be higher or lower than the average cost that NMFS reported in the DEIS and that is why we recommended that NMFS represent this uncertainty by presenting a range of possible costs in its economic analysis. We did not recommend applying higher gear loss and replacement rates to the entire fishing industry as NOAA seems to suggest in its comments. Therefore, we did not revise the report in response to this comment.

11. We recognize that portions of Maine's state waters are proposed to be exempt from the changes to the ALWTR plan. This does not change the fact that NMFS's gear research team estimated that gear loss would vary by area of fishing operation and that, according to the MLA, NMFS's estimates are likely to be too low in Maine's rocky bottom areas that will be subject to the new regulation. Furthermore,

the report does not attempt to identify a particular level of uncertainty related to gear loss as NOAA contends. For these reasons, we did not revise the report in response to this comment.

12. We do not agree with NOAA's comment that markings for identifying sinking groundline and gear in exempted areas are not feasible or practical at this time. In the DEIS, NOAA proposed requiring that vertical line be marked. If such marking is feasible and practical for vertical line, the same type of marking should be feasible and practical for sinking groundline. Many scientists we spoke to indicated that sinking groundline should be marked. Consequently, we did not revise the report in response to this comment.

13. Because the draft report already included a paragraph which discusses the status of efforts to use "chip technology" to identify fishing gear, including that NMFS believes that it is not yet ready to be implemented, we made no changes in response to this comment.

14. If NOAA is unable to complete its strategy for assessing industry compliance prior to finalizing its proposed regulations, we believe the agency should have the strategy in place by the effective date of the final regulations so that it is in a position to evaluate the effectiveness of its regulatory changes from their inception. We did not revise the report in response to this comment.

End Notes

[1] This chapter addresses the western North Atlantic stock of right whales, the Gulf of Maine stock of humpback whales, and the western North Atlantic stock of fin whales. NMFS is an agency of the Department of Commerce's National Oceanic and Atmospheric Administration.

[2] Traps are also referred to as pots.

[3] There are many different types of bottom-dwelling Atlantic groundfish, including haddock, cod, and various flounder.

[4] In this chapter, we will refer to the Atlantic Large Whale Take Reduction Plan as the ALWTR plan.

[5] NMFS. *Draft Environmental Impact Statement for Amending the Atlantic Large Whale Take Reduction Plan: Broad-Based Gear Modifications.* (Washington, D.C.: February 2005).

[6] Sinking groundline is also referred to as neutrally buoyant groundline.

[7] 70 *Fed. Reg.* 35893 (June 21, 2005).

[8] NMFS has authorized the Provincetown Center for Coastal Studies as the lead organization on the east coast to disentangle large whales.

[9] Caswell, H.; Fujiwara, M.; Brault, S. "Declining survival threatens the North Atlantic right whale," *Proceedings of the National Academy of Sciences,* vol. 96, no. 6 (1999).

[10] Fin whales were rarely hunted in U.S. waters, except near the shores of Provincetown, Massachusetts in the late 1800s.

[11] Copepods are small crustaceans.

[12] NOAA delegated its MMPA responsibilities to NMFS.

[13] 16 U.S.C. § 1387.

[14] As defined in the MMPA, the term "take" means to harass, hunt, capture, or kill or to attempt to harass, hunt, capture or kill a marine mammal. 16 U.S.C. § 1362(13). Take reduction plans must be developed to mitigate the effects of Category I and II fisheries, defined as fisheries that have frequent incidental mortality and serious injury of marine mammals (Category I) and fisheries that have occasional incidental mortality and serious injury of marine mammals (Category II). 16 U.S.C. § 1387(c)(1)(A).

[15] The Magnuson Fishery Conservation and Management Act of 1976 (since renamed the Magnuson-Stevens Act) created eight regional councils to manage fishery resources within federal waters (from 3 to 200 miles off the coast).

[16] As defined in the MMPA, potential biological removal is the maximum number of animals, not including natural mortalities that may be removed from a marine mammal stock annually while allowing that stock to reach or maintain its optimal sustainable population.

[17] The MMPA does not define "insignificant" mortality and serious injury rates approaching zero. NMFS has established a "zero mortality rate goal" as no more than 10 percent of the potential biological removal level for each stock.

[18] The Department of Interior administers the ESA for freshwater and land species and the Department of Commerce through NMFS administers the act for marine species.

[19] A fisherman with a multispecies permit is able to target more than one species of groundfish, such as haddock, yellowtail flounder, winter flounder, Atlantic cod, white hake, and American plaice.

[20] Additional fisheries include black sea bass, scup, conch/whelk, shrimp, red crab, hagfish, Jonah crab, Northeast driftnet and Northeast anchored float gillnet.

[21] Waring, G.T; Josephson, E.; Fairfied, C.P.; Maze-Foley, K. *U.S. Atlantic and Gulf of Mexico Marine Mammal Stock Assessments—2006*. (Woods Hole, MA: 2007).

[22] The stock assessment report uses the term "potential biological removal" to refer to the maximum number of human-related mortalities that can occur annually while allowing a stock to reach or maintain its optimum sustainable population.

[23] Waring, G.T; Pace, R.M.; Quintal, J.M.; Fairfied, C.P.; Maze-Foley, K. *U.S. Atlantic and Gulf of Mexico Marine Mammal Stock Assessments—2003*. (Woods Hole, MA: 2004).

[24] NMFS is required under the MMPA to prepare stock assessment reports of marine mammals, including large whales, in order to monitor their population status. 16 U.S.C. § 1386.

[25] Waring, G.T; Josephson, E.; Fairfied, C.P.; Maze-Foley, K. *U.S. Atlantic and Gulf of Mexico Marine Mammal Stock Assessments—2006*. (Woods Hole, MA: 2007).

[26] While NMFS can develop regulations in response to recovering gear of unknown origin from entangled whales, according to a NMFS official, all the regulatory actions the agency has taken in the past have been in response to entanglements in U.S. gear, or gear consistent with U.S. fisheries.

[27] The New England Aquarium maintains a photo identification database, funded by NMFS, which includes all known photographed sightings of right whales since 1935. NMFS's aerial surveys, research groups, whale watch vessels, and others have contributed to the database.

[28] Knowlton, A.R.; Marx, M.K.; Pettis, H.M.; Hamilton, P.K.; Kraus, S.D. Analysis of Scarring on North Atlantic Right Whales (Eubalaena glacialis): Monitoring Rates of Entanglement Interaction 1980-2002. Final report to NMFS submitted by The New England Aquarium (2005).

[29] Robbins, J. and Mattila, D. *Estimating Humpback Whale (Megaptera novaeangliae) Entanglement Rates on the Basis of Scar Evidence*. Report to the Northeast Fisheries Science Center submitted by The Center for Coastal Studies (2004).

[30] When whales are discovered entangled, NMFS sends a team that may attempt to disentangle the whale, depending on the nature of the entanglement. The team attempts to obtain

information about the gear involved in the entanglement, such as whether it is vertical line or groundline.

[31] Johnson, A.; Salvador, G.; Kenney, J.; Robbins, J.; Kraus, S.; Landry, S.; Clapham, P. "Fishing Gear Involved in Entanglements of Right and Humpback Whales," *Marine Mammal Science*, vol. 21, no. 4, (2005).

[32] The North Atlantic Right Whale Consortium Sighting Database, maintained by the University of Rhode Island, includes sightings from NMFS's aerial survey as well as other sightings along the eastern seaboard. It does not include photographs, like the database maintained by the New England Aquarium, as researchers and others are not able to photograph each whale that is spotted.

[33] Mackintosh, W. and Wiley, D. *The Development and Operational Testing of Gillnet Fishing Gear Equipped with Five 600 lb Weak Links*. Report to the International Wildlife Coalition and the Northeast Consortium. (May 6, 2005).

[34] NMFS opened a formal public comment period on the DEIS during which any stakeholder could submit comments.

[35] In addition to the field testing of sinking groundline with fishermen described in this chapter, NMFS conducted additional testing of sinking groundline, including using a line testing machine that simulates some of the long-term wear and tear that lines experience in the field. NMFS gear specialists are former fishermen or boat captains who test fishing gear and conduct outreach with fishermen.

[36] While the formal survey was conducted in 2003, NMFS gear specialists told us that they interviewed fishermen throughout the duration of the test and still informally check in with fishermen who continue to use the line today.

[37] Experimental sinking groundline refers to rope that is under development. It is being tested because it is made with different materials or coatings than rope currently on the market.

[38] The Consortium for Wildlife Bycatch Reduction is a NMFS-funded partnership between scientists and industry to design, field test, and promote fishing gear that minimizes incidental harm to marine life. Founding members include the University of New Hampshire, Duke University , the Maine , and the New England Aquarium.

[39] NMFS's estimate for the lifespan of 3/8" sinking groundline is 6 years.

[40] NMFS's analysis was based on vessels, but in this chapter we will refer to vessels as fishermen because they are affected by the regulation and would incur the costs.

[41] NMFS estimated the total cost to the fishing industry from the gear modifications outlined in each of its six proposed alternatives to revising the ALWTR plan in 2003 dollars. In this chapter, we discuss the costs of the two preferred alternatives, both of which NMFS estimated would cost the fishing industry about $14 million annually.

[42] Lifespan percentages are GAO's analysis of NMFS lifespan data.

[43] The prices reported for rope in the DEIS were adjusted to 2007 dollars to account for inflation.

[44] Federal waters, that is, waters under the jurisdiction of the United States, extend from 3 nautical miles to 200 nautical miles offshore. State waters extend from the shore to 3 nautical miles offshore.

[45] State-permitted fishermen are those that operate in state waters and are required to obtain a permit from the state.

[46] This estimate only includes state-permitted fishermen, not those that may also have a federal permit.

[47] NOAA guidance, which NMFS followed to conduct the economic assessment within the DEIS, does not require the agency to estimate changes in regional employment and income.

[48] NMFS identified over 100 vessels within each county to determine if they would be at-risk of being affected; however, in this chapter we refer to them as fishermen.

[49] The Magnuson-Stevens Act requires that NMFS consider impacts on "communities." The act defines "fishing community" as "a community which is substantially dependent on or substantially engaged in the harvest or processing of fishery resources to meet social and economic needs and includes fishing vessel owners, operators, and crew and United States

fish processors that are based in such community." 16 U.S.C. § 1802(16). NMFS used counties as a proxy for communities because fishermen may reside in an area different from where they port their vessel. In addition, much of the landings data was county specific.

[50] For vertical lines that are less than 60 feet, fishermen would be required to place one 4-inch mark in the center of the line.

[51] The Whale Center of New England is a nonprofit organization that conducts research, conservation, and education.

[52] The lack of federal funding in 2006 prevented Maine from conducting the survey that year, but the state plans to resume the survey in 2007.

In: Marine Mammal Protection Issues
Editors: Derek L. Caruana

ISBN: 978-1-60741-540-4
© 2010 Nova Science Publishers, Inc.

Chapter 3

NATIONAL MARINE FISHERIES SERVICE: IMPROVEMENTS ARE NEEDED IN THE FEDERAL PROCESS USED TO PROTECT MARINE MAMMALS FROM COMMERCIAL FISHING

United States Government Accountability Office

WHY GAO DID THIS STUDY

Because marine mammals, such as whales and dolphins, often inhabit waters where commercial fishing occurs, they can become entangled in fishing gear, which may injure or kill them—this is referred to as "incidental take." The 1994 amendments to the Marine Mammal Protection Act (MMPA) require the National Marine Fisheries Service (NMFS) to establish take reduction teams for certain marine mammals to develop measures to reduce their incidental takes. GAO was asked to determine the extent to which NMFS (1) can accurately identify the marine mammal stocks—generally a population of animals of the same species located in a common area—that meet the MMPA's requirements for establishing such teams, (2) has established teams for those stocks that meet the requirements, (3) has met the MMPA's deadlines for the teams subject to them, and (4) evaluates the effectiveness of take reduction regulations. GAO reviewed the MMPA, and NMFS data on marine mammals, and take reduction team documents and obtained the views of NMFS officials, scientists, and take reduction team members.

WHAT GAO RECOMMENDS

GAO is proposing matters for congressional consideration, including requiring NMFS to report on the data, resource, and other limitations that prevent it from meeting the MMPA's requirements for take reduction teams; and recommending that NMFS develop a comprehensive strategy for assessing plan effectiveness. The agency agreed with our recommendation to develop such a strategy.

WHAT GAO FOUND

Significant limitations in available data make it difficult for NMFS to accurately determine which marine mammal stocks meet the statutory requirements for establishing take reduction teams. For most stocks, NMFS relies on incomplete, outdated, or imprecise data on stocks' population size or mortality to calculate the extent of incidental take. As a result, the agency may overlook some marine mammal stocks that meet the MMPA's requirements for establishing teams or inappropriately identify others as meeting them. NMFS officials told GAO they are aware of the data limitations but lack funding to implement their plans to improve the data.

On the basis of NMFS's available information, GAO identified 30 marine mammal stocks that have met the MMPA's requirements for establishing a take reduction team, and NMFS has established six teams that cover 16 of them. For the other 14 stocks, the agency has not complied with the MMPA's requirements. For example, false killer whales, found off the Hawaiian Islands, have met the statutory requirements since 2004, but NMFS has not established a team for them because, according to NMFS officials, the agency lacks sufficient funds. NMFS officials told GAO that the agency has not established teams for the other stocks that meet the MMPA's requirements for reasons such as the following: (1) data on these stocks are outdated or incomplete, and the agency lacks funds to obtain better information and (2) causes other than fishery-related incidental take, such as sonar used by the U.S. Navy, may contribute to their injury or death, therefore changes to fishing practices would not solve the problem.

For the five take reduction teams subject to the MMPA's deadlines, the agency has had limited success in meeting the deadlines for establishing teams, developing draft take reduction plans, and publishing proposed and final plans and regulations to implement them. For example, NMFS established three of the five teams—the Atlantic Large Whale, Pelagic Longline, and Bottlenose Dolphin—from 3 months to over 5 years past the deadline. NMFS officials attributed the delays in establishing one of the teams to a lack of information about stock population size and mortality, which teams need to consider before developing draft take reduction plans.

NMFS does not have a comprehensive strategy for assessing the effectiveness of take reduction plans and implementing regulations that have been implemented. NMFS has taken some steps to define goals, monitor compliance, and assess whether the goals have been met, but shortcomings in its approach and limitations in its performance data weaken its ability to assess the success of its take reduction regulations. For example, without adequate information about compliance, if incidental takes continue once the regulations have been implemented, it will be difficult to determine whether the regulations were ineffective or whether the fisheries were not complying with them.

ABBREVIATIONS

CV	coefficient of variation
ESA	Endangered Species Act
MMPA	Marine Mammal Protection Act
NMFS	National Marine Fisheries Service
NOAA	National Oceanic and Atmospheric Administration
OMB	Office of Management and Budget

December 8, 2008

The Honorable Nick J. Rahall II
Chairman Committee on Natural Resources
House of Representatives
Dear Mr. Chairman:

Marine mammals—such as whales, dolphins, and porpoises—often swim or feed in waters where commercial fishing occurs and can become entangled in fishing gear, which may seriously injure or kill them—this is referred to as "incidental take." The National Oceanic and Atmospheric Administration's (NOAA) National Marine Fisheries Service (NMFS) estimates that commercial fishing operations result in thousands of such incidental takes each year. For example, large whales, such as the North Atlantic right whale, may become entangled in the lines that connect lobster traps; smaller pilot whales may become entangled in longline fishing gear used to catch fish such as tuna or swordfish; and dolphins and porpoises may become entangled in commercial fishing nets used to catch sardines, salmon, or cod. For at least five marine mammals, incidental takes as a result of commercial fishing operations are occurring at unsustainable levels.[1]

The 1994 amendments to the Marine Mammal Protection Act (MMPA) of 1972 require NMFS to establish take reduction teams to develop regulatory or voluntary measures for the reduction of incidental mortality and serious injury to marine mammals during the course of commercial fishing operations. Under the MMPA, NMFS, in general, must establish take reduction teams for each marine mammal strategic stock that interacts with a Category I or Category II commercial fishery.[2] These key terms are defined as follows:

- A commercial fishery is a group of fishermen who use similar gear to catch the same types of fish, in a common geographic area, and then sell them.
- A Category I fishery is a commercial fishery that has frequent incidental takes of marine mammals, while a Category II fishery has occasional incidental takes.
- A stock is a group of marine mammals of the same species located in a common spatial arrangement that interbreed when mature.

The MMPA defines a marine mammal stock as strategic, if it (1) is listed as threatened or endangered under the Endangered Species Act (ESA), (2) is declining and likely to be listed as a threatened species under the ESA within the foreseeable future, (3) is designated as depleted under the MMPA, or (4) has a level of direct human-caused mortality and serious injury that exceeds the stock's potential biological removal level. In this chapter we use the term "maximum removal level," rather than potential biological removal level; this term is defined as the maximum number of animals, not including natural mortalities, that may be removed from a marine mammal stock while allowing that stock to reach or maintain its optimum sustainable population.[3]

NMFS periodically surveys marine mammal populations to determine whether they are growing, remaining stable, or declining, so that it can calculate the maximum removal level. The results of this research are published in annual stock assessment reports for 156 stocks that fall under NMFS's jurisdiction. Additionally, NMFS annually publishes lists that classify commercial fisheries as Category I, II, or III based on data such as those from observers who are placed on boats, logbooks that are kept by fishermen, and data gathered by scientists at universities and government agencies, among others, documenting instances where marine mammals are found stranded or dead as a result of fishing or other human causes.[4]

According to the MMPA, take reduction teams must include representatives from the commercial fishing industry, Regional Fishery Management Councils, interstate fisheries commissions, environmental groups, academic and scientific organizations, and state and federal governments. Once NMFS has established a team, the members meet, review the available information regarding marine mammal takes and fisheries interactions, and develop draft take reduction plans. The plans recommend regulatory and voluntary measures, such as modifications in fishing gear or practices that should reduce serious injury or mortality of marine mammals caused by commercial fishing. As specified in the MMPA, NMFS then translates these draft plans into final take reduction plans and implementing regulations.

The 1994 amendments to the MMPA set several deadlines for establishing take reduction teams, developing draft take reduction plans, and publishing proposed and final plans and implementing regulations. Specifically,

- NMFS must establish a take reduction team within 30 days after a final stock assessment report indicates that a stock is strategic and it is listed in the current list of fisheries as interacting with a Category I or II fishery.
- Take reduction team members must develop a draft plan and submit the plan to NMFS within 6 months after the take reduction team is established.[5]
- NMFS must publish a proposed take reduction plan in the *Federal Register* within 60 days of receiving the team's draft plan.[6]
- NMFS must hold a public comment period on the proposed take reduction plan for up to 90 days after its publication.

- NMFS must publish a final take reduction plan in the *Federal Register* 60 days after the public comment period on the proposed plan ends.

As Congress prepares to consider the reauthorization of the MMPA, you asked us to determine the extent to which (1) available data allow NMFS to accurately identify the marine mammal stocks that meet the MMPA's requirements for establishing take reduction teams; (2) NMFS has established take reduction teams for those marine mammal stocks that meet the statutory requirements; (3) NMFS has met the statutory deadlines established in the MMPA for the take reduction teams subject to the deadlines, and the reasons for any delays; and (4) NMFS has developed a comprehensive strategy for evaluating the effectiveness of the take reduction plans that have been implemented.

To determine the extent to which available data allow NMFS to identify the marine mammal stocks that meet the MMPA's requirements for establishing take reduction teams, we identified several key data elements, such as human-caused mortality estimates and maximum removal levels, that NMFS uses to determine whether a marine mammal stock meets the statutory requirements for establishing a take reduction team, as well as the criteria NMFS uses to assess data quality for these key data elements.

We then reviewed all of the 2007 marine mammal stock assessment reports and analyzed reports for the 113 stocks not currently covered by take reduction teams and not listed as threatened or endangered species or designated as depleted species. After removing those that lacked human-caused mortality estimates or maximum removal levels, we reviewed a sample of the remaining 74 stocks that did have these determinations to assess the reliability of the information used to determine human-caused mortality estimates and the maximum removal levels. We reviewed a sample of the reports for these stocks to identify any data uncertainties that may limit NMFS's ability to accurately identify stocks that meet the statutory requirements for establishing take reduction teams. To determine the extent to which NMFS has established take reduction teams, we analyzed stock assessment reports and lists of fisheries from 1995 through 2008 to identify marine mammal stocks that meet the statutory requirements but are not currently covered by a team. Additionally, we interviewed NMFS officials and obtained documentation on the stocks for which the agency has established take reduction teams. To identify the extent to which NMFS has met the deadlines established in the MMPA, we identified the deadlines listed in the

MMPA for take reduction teams and obtained documentation, such as take reduction plans and implementing regulations, to determine whether NMFS met the statutory deadlines. To identify the reasons for any delays in meeting the statutory deadlines, we interviewed the NMFS staff members that coordinate the take reduction teams, staff from NOAA's Office of General Counsel, marine biologists in NMFS's Fishery Science Centers, and members of each of the five take reduction teams subject to the MMPA's deadlines.[7] To identify the extent to which NMFS has developed a comprehensive strategy for evaluating the effectiveness of its take reduction plans, we reviewed data from stock assessment reports on the level of fishery-related mortality and serious injury as well as maximum removal levels before and after a plan's implementation. We also interviewed NMFS officials, including agency staff that coordinate take reduction teams, regarding how they assess the effectiveness of take reduction plans. We conducted this performance audit from September 2007 to December 2008 in accordance with generally accepted government auditing standards. Those standards require that we plan and perform the audit to obtain sufficient, appropriate evidence to provide a reasonable basis for our findings and conclusions based on our audit objectives. We believe that the evidence obtained provides a reasonable basis for our findings and conclusions based on our audit objectives. Appendix I provides additional detail on our scope and methodology.

RESULTS IN BRIEF

Significant limitations in available information make it difficult for NMFS to accurately determine which marine mammal stocks meet the statutory requirements for establishing take reduction teams. For most stocks, NMFS relies on incomplete, outdated, or imprecise information about human-caused mortality or the maximum removal level to calculate whether incidental take is above acceptable levels and thereby determine if the stocks meet the MMPA's definition of "strategic"—one of two triggers for establishing a take reduction team. For example, our review of stock assessment reports found that NMFS did not have a human-caused mortality estimate or a maximum removal level for 39 of 113 marine mammal stocks, making it impossible to determine their strategic status in accordance with the MMPA's requirements. For the remaining 74 stocks, NMFS has some data to determine whether incidental takes exceeded acceptable levels, but these data have significant limitations that call into question their accuracy. Specifically, for an estimated 11 of the 74 stocks, the maximum removal levels were based on information that was 8 years old or older. Marine mammal research has shown that using such outdated data increases the possibility that significant population declines could have occurred of which NMFS is unaware. In addition, our review of a sample of stock assessment reports from 2007 frequently found that NMFS used population size and fishery-related mortality estimates that were less precise than recommended by the agency's own guidance, decreasing the likelihood that strategic status determinations based on this information are accurate. Relying on incomplete, outdated, or imprecise information about human-caused mortality and maximum removal levels may lead NMFS to overlook some marine mammal stocks that meet the statutory requirements for establishing take reduction teams or inappropriately identify others as meeting them. NMFS officials told us that funding constraints limit their ability to gather sufficient data, although the agency has taken some steps to identify its data needs. For example, in a 2004 study, NMFS identified the actions and resources needed to improve the marine mammal stock assessment data that support MMPA decisions; however, officials told us that they have not received the resources necessary to complete the actions identified in the report.

On the basis of NMFS's available information, we identified 30 marine mammal stocks that met the MMPA's requirement for establishing a take reduction team, and NMFS has established six teams that cover 16 of them. For the 14 other marine mammal stocks for which the agency's available information shows them to be strategic and interacting with Category I or II fisheries, NMFS has not complied with the MMPA's requirement to establish take reduction teams and, in some cases, has not been in compliance for several years. NMFS officials told us that the agency has not established teams for these 14 marine mammal stocks for various reasons. First, the agency lacked sufficient funds to establish a team for one marine mammal stock—the Hawaiian stock of false killer whales—that has met the statutory requirements since 2004. Second, for 8 of the 14 stocks, NMFS information about the stocks' population size or mortality is outdated or incomplete, and the agency lacks funds to obtain better information. Third, for 4 of the 14 stocks, commercial fisheries account for few or no incidental takes, and other causes, such as acoustic activities, for example, sonar used by the U.S. Navy, may contribute to the serious injury and mortality of some of these stocks, so establishing teams for them would not be appropriate. Finally, the population size of one marine mammal stock—the Central North Pacific stock of humpback whales—is increasing; therefore establishing a team for this stock is a low priority.

For the five take reduction teams subject to the MMPA's deadlines, NMFS has had limited success in meeting them for various reasons. Specifically,

- NMFS missed statutory deadlines for establishing three of the five teams—the Atlantic Large Whale, Pelagic Longline, and Bottlenose Dolphin—by 3 months to more than 5 years. According to NMFS officials, the reason for delays in establishing one of these three teams was a lack of information, such as information on stocks' population size and mortality, that team members need to consider before developing draft take reduction plans.

- Two of the five teams did not submit their draft take reduction plans to NMFS within the statutory deadlines. In one case the team missed the deadline because it had difficulty reaching consensus on a plan, and in the other case there was an unexpected death of a key team member 1 week before the plan was due.

- NMFS did not publish proposed take reduction plans in accordance with the statutory deadlines for the five teams. According to agency officials, these deadlines were missed because of the time needed to complete the federal rulemaking process, among other things. However, NMFS complied with the statutory deadline for the public comment periods for the five teams that have reached this stage of the process.

- NMFS missed the statutory deadline for publishing final take reduction plans and implementing regulations for four of the five teams—the Atlantic Large Whale, Pacific Offshore Cetacean, Bottlenose Dolphin, and the Pelagic Longline—by 8 days to over a year. NMFS attributed the delays to the time necessary to respond to the public comments it received on the proposed plan before it could publish the final plan, among other things.

NMFS does not have a comprehensive strategy for assessing the effectiveness of take reduction plans and implementing regulations once they have been implemented. The Government Performance and Results Act of 1993 provides a foundation for examining agency performance goals and results. Our work related to the act and the experience of leading organizations have shown the importance of developing a strategy for assessing performance that includes, among other things, program performance goals that identify the desired results of program activities and reliable information that can be used to assess results. In the context of NMFS's efforts to measure the success of the regulations that result from take reduction plans, we believe such a strategy would include, at a minimum, (1) performance goals that identify the desired outcomes of the take reduction regulations; (2) steps for assessing the effectiveness of potential take reduction regulations, such as fishing gear modifications, in achieving the goals; (3) a process for monitoring the fishing industry's compliance with the requirements of the take reduction regulations; and (4) reliable data assessing the regulations' effect on achieving the goals. Instead of such a comprehensive strategy, we found that although NMFS uses short- and long-term goals established by the MMPA to evaluate the success of take reduction regulations, these goals and the data that NMFS uses to measure the impact of the take reduction regulations have a number of limitations. For example, according to officials we spoke with, it is difficult to assess the impact of the regulations in a 6- month period, as required by the MMPA's short-term goal. In addition, while NMFS has taken steps to identify the impact of proposed take reduction regulations prior to their implementation, the agency has limited information about the fishing industry's compliance with the regulations once they have been implemented. As a result, when incidental takes occur in fisheries covered by take reduction regulations, it is difficult for NMFS to determine whether the regulations were not effective in meeting the MMPA's goals or whether the fisheries were not complying with the regulations.

To facilitate the oversight of NMFS's progress and capacity to meet the statutory requirements under the MMPA for take reduction teams, Congress may wish to consider (1) directing NMFS to report on the key factors that affect its ability to meet the MMPA's requirements for establishing teams and meeting statutory deadlines, including data, resources, or other limitations; (2) amending the statutory requirements in the MMPA for establishing a take reduction team to stipulate that not only must a marine mammal stock be strategic and interacting with a Category I or II fishery but that the fishery with which the marine mammal stock interacts causes at least occasional incidental mortality or serious injury of that particular marine mammal stock; and, (3) amending the MMPA to ensure that the statutory deadlines give NMFS adequate time to complete take reduction plans and implementing regulations. We are also recommending that NMFS develop a comprehensive strategy for assessing the effectiveness of each take reduction plan and implementing regulations. In its comments on a draft of this chapter NOAA agreed with our recommendation to develop such a comprehensive strategy.

BACKGROUND

The MMPA was enacted in 1972 to ensure that marine mammals are maintained at or restored to healthy population levels. Among other things, this act established the Marine Mammal Commission, which must continually review the condition of marine mammal stocks and recommend to the appropriate federal officials and Congress any steps it deems necessary or desirable for the protection and conservation of marine mammals.[8] In 1994, the MMPA was amended to create a process for establishing take reduction teams to manage incidental takes—serious injury or death—in the course of commercial fishing operations. Commercial fishing in areas where marine mammals swim, feed, or breed is considered one of the main human causes of incidental take. Marine mammals can become entangled in fishing equipment such as nets or hooks, although specific threats vary by the fishing techniques used. Appendix II provides details on commercial fishing techniques that can result in incidental take, including gillnetting, longlining, trap/pot fishing, and trawling, as well as examples of the marine mammals affected.

Under the 1994 amendments to the MMPA, NMFS must establish take reduction teams when two requirements are satisfied: (1) NMFS designates the stock as strategic in a final stock assessment report, and (2) the stock interacts with a commercial fishery listed as Category I or II in the current list of fisheries.[9] According to the MMPA, if there is insufficient funding to develop and implement take reduction plans for all stocks that meet the requirements, NMFS should establish teams based on specified priorities.[10] For the majority of stocks, NMFS determines strategic status by comparing whether human-caused mortality exceeds the maximum removal level (see fig. 1).[11] Human-caused mortality and serious injury (hereafter known as human-caused mortality) is estimated by adding fishery-related mortality estimates to mortality caused by other human sources, as follows:

- Fishery-related mortality and serious injury estimates (hereafter known as fishery-related mortality estimates) are generated based on data from NMFS's fishery observer programs, whereby individuals board commercial fishing vessels and document instances of incidental take. NMFS also uses anecdotal information from scientists, fishermen, and others about additional incidental take to make these estimates.

- Mortality and serious injury caused by other human sources such as collisions with large ships or authorized subsistence hunting of marine mammals by Alaska natives.

- The maximum removal level—technically known as the potential biological removal level—is calculated for each marine mammal stock by multiplying three factors:

- The minimum population estimate (hereafter known as the population size estimate) for the specific stock of marine mammals.[12]

- Two adjustments designed to (1) factor in the expected rate of natural growth for a stock and (2) reduce the risks associated with data uncertainties, especially for stocks listed as endangered or threatened or designated as depleted. By altering the values of these adjustments, NMFS can make the maximum removal level more conservative— meaning that fewer incidental takes will be allowed—in cases of uncertain data, and therefore make it less likely that they will identify a stock as nonstrategic.

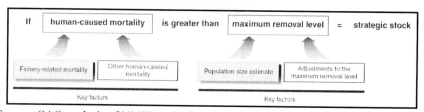

Source: GAO analysis of NMF's guidelines for assessing marine mammal stocks.

Figure 1. Determining Strategic Status by Comparing Human-Caused Mortality to the Maximum Removal Level.

The MMPA requires NMFS to assess the status of each stock under its jurisdiction and determine whether it is strategic or not. NMFS publishes annual stock assessment reports that include, among other things, the strategic status of each marine mammal stock and the information used to make these strategic status determinations. Information contained in the reports must be based on the best scientific information available. NMFS's Fishery Science Centers are responsible for publishing the stock assessment reports, and the Office of Protected Resources, along with NMFS regional offices, is responsible for using the data from the reports to decide whether to establish a take reduction team. Regional Scientific Review Groups—composed of individuals with expertise in marine mammal biology, commercial fishing technology and practices, and other areas—review all stock assessment reports prior to publication.[13] NMFS also uses fishery-related mortality estimates and maximum removal levels in the stock assessment reports to categorize fisheries in its annual list of fisheries.[14] Under the amended MMPA, commercial fisheries are classified as Category I if they have frequent incidental take of marine mammals and as Category II if they have occasional take.[15]

Once a stock is identified as requiring a take reduction team—because it is strategic and interacts with a Category I or II fishery—the MMPA requires NMFS to establish a team and appoint take reduction team members. The MMPA requires the take reduction team members to develop and submit a draft take reduction plan designed to reduce the incidental take of marine mammals by commercial fishing operations. If NMFS lacks sufficient funding to develop and implement a take reduction plan for all stocks that satisfy the MMPA's requirements, the MMPA directs NMFS to give highest priority to take reduction plans for those stocks (1) for which incidental mortality and serious injury exceed the maximum removal level, (2) with a small population size, and (3) that are declining most rapidly. The MMPA requires that draft take reduction plans be developed by consensus among take reduction team members. If take reduction team members cannot reach consensus, the members must submit the range of possibilities they considered and the views of both the majority and minority to NMFS. These draft plans may include regulatory measures (known as take reduction regulations) such as gear modifications or geographical area closures that fisheries would be required to follow and voluntary measures such as research plans for identifying the primary causes and solutions for incidental take or education and outreach for commercial fishermen.

After the take reduction team members develop and submit a draft take reduction plan to NMFS, the agency must publish a proposed plan in the *Federal Register*. The MMPA requires NMFS to take the team's draft plan into consideration when it develops a proposed plan but does not require adoption of the draft plan.[16] If the team fails to meet its deadline for submitting a draft plan to NMFS, the MMPA requires NMFS to develop and propose a plan on its own. For strategic stocks, the proposed plan must include measures NMFS expects will reduce incidental take below the maximum removal level within 6 months of the plan's implementation. Once the proposed plan is published in the *Federal Register*, NMFS must solicit public comments on the plan before the agency finalizes and implements it by publishing a final plan in the *Federal Register*. NMFS's development and publication of proposed and final plans are subject to several laws, including the following:

- **Endangered Species Act:** The act requires consultation among federal agencies including NMFS and the U.S. Fish and Wildlife Service to ensure that any take reduction plan is not likely to jeopardize the continued existence of any endangered or threatened species.

- **National Environmental Policy Act:** The act requires NMFS to evaluate the likely environmental effects of any take reduction plan using an environmental assessment or, if the plans will likely have significant environmental effects, a more detailed environmental impact statement.

- **Regulatory Flexibility Act:** The act requires NMFS to assess the economic impact of any take reduction plan on small entities.[17]

The proposed and final take reduction plans are also subject to the requirements of the Coastal Zone Management Act, Information Quality Act, Magnuson-Stevens Act, and the Paperwork Reduction Act, among others. In addition to these laws, the proposed and final take reduction plans are subject to the requirements of four executive orders.[18] For example, one executive order requires NMFS to submit the proposed and final take reduction plans to the Office of Management and Budget (OMB) for review if NMFS or OMB determines that the plan is a significant regulatory action.

The 1994 amendments to the MMPA provide deadlines to establish take reduction teams and develop and publish proposed and final plans. Table 1 outlines these statutory requirements and deadlines.

Table 1. MMPA's Take Reduction Team Requirements and Deadlines

Requirement	Deadline
NMFS establishes take reduction team	30 days after a final stock assessment report indicates that a stock is strategic and the current list of fisheries lists the stock as interacting with a Category I or II fishery

Take reduction team members develop a draft plan and submit the draft plan to NMFS	6 months after take reduction team is established[a]
NMFS translates draft plan into a proposed take reduction plan and implementing regulations and publishes them in the *Federal Register*	60 days after draft plan is submitted[b]
NMFS holds a public comment period on the proposed take reduction plan and implementing regulations	Up to 90 days after proposed plan's publication
NMFS publishes a final take reduction plan and implementing regulations in the *Federal Register*	60 days after closure of the public comment period on the proposed plan

Source: GAO analysis of the 1994 amendments to the MMPA.

[a] If the team's stocks have human-caused mortality and serious injury below the maximum removal level and interact with a Category I or II fishery, this deadline is 11 months instead of 6 months.

[b] If the team does not meet its submission deadline for the draft plan, NMFS must publish a proposed plan 8 months after the team's establishment for strategic stocks whose human-caused mortality and serious injury are above the maximum removal level. For strategic stocks whose human-caused mortality and serious injury are below the maximum removal level, the deadline is 13 months after the team's establishment.

NMFS FACES SIGNIFICANT CHALLENGES IN ACCURATELY IDENTIFYING MARINE MAMMAL STOCKS THAT MEET THE STATUTORY REQUIREMENTS FOR ESTABLISHING TAKE REDUCTION TEAMS BECAUSE OF DATA LIMITATIONS

Significant limitations in available information make it difficult for NMFS to accurately determine which marine mammal stocks meet the statutory requirements for establishing take reduction teams. The MMPA states that stocks are strategic—one of two triggers for establishing a take reduction team—if their human-caused mortality exceeds maximum removal levels. However, the information NMFS uses to calculate human-caused mortality or the maximum removal level for most stocks is incomplete, outdated, or imprecise, a fact that may lead NMFS to overlook some marine mammal stocks that meet the statutory requirements for establishing take reduction teams and inappropriately identify others as meeting them. NMFS officials said that funding constraints limit their ability to gather sufficient data, although the agency has taken steps to identify its data needs.

Incomplete Information Reduces the Reliability of NMFS's Strategic Status Determinations

Our review of stock assessment reports from 2007 found that NMFS was missing key information to make well-informed strategic status determinations for a significant number of marine mammal stocks. According to the MMPA, a stock is designated strategic—one of two triggers for establishing a take reduction team—if the human-caused mortality estimate exceeds the maximum removal level.[19] Our review of stock assessment reports from 2007 found that 39 of 113 stocks are either missing human-caused mortality estimates or maximum removal levels, making it impossible to determine strategic status in accordance with the MMPA requirements.[20] As a result, for these 39 stocks, NMFS is determining strategic status without key information and therefore might not accurately determine whether a stock requires a take reduction team. According to NMFS officials, maximum removal level and human-caused mortality estimates are often missing because scientists have been unable to gather the necessary data to make these determinations.

In the absence of human-caused mortality estimates or maximum removal levels, NMFS must make more subjective—and potentially inaccurate— strategic status determinations for some marine mammal stocks. In these cases, NMFS guidance directs scientists to use professional judgment to determine whether a stock is strategic. According to NMFS officials, scientists may use a variety of sources to make these decisions, including scientists' field observations of the marine mammals. However, Marine Mammal Commission officials we spoke with stated that decisions based on professional judgment are less certain than those based on data about human-caused mortality and maximum removal levels and could result in some marine mammal stocks that should be identified as strategic not being identified as such.

Even in cases where the stock assessment reports include human-caused mortality estimates and maximum removal levels for a stock, the human-caused mortality estimates may be inaccurate because the information on which they are based is incomplete. Human-caused mortality estimates are based in part on fishery-related mortality. However, according to Marine Mammal Commission officials, in some cases, mortality may be occurring in fisheries where NMFS does not systematically collect mortality information. Specifically, NMFS's observer programs—a key source of data NMFS uses to calculate fishery-related mortality estimates-—gather information for only half of the total fisheries, but incidental take may also be occurring in some fisheries that are not observed, especially those that are classified as Category I or II. Observer program officials told us that funding limitations prohibit coverage of all Category I or II fisheries.

In addition, our review of 2007 stock assessment reports found instances where fishery-related mortality estimates were missing important information. For example, NMFS scientists identified spinner and bottlenose dolphins in Hawaii as nonstrategic, but raised concerns about these decisions because the estimates of fishery-related mortality for the stocks were likely to be incomplete. Specifically, they stated that while the agency has observer program data showing that incidental take from a longline fishery was below the maximum removal level, it did not have observer programs in gillnet fisheries that were also likely to incidentally take the stocks, and therefore might have increased the fishery-related mortality estimate if these fisheries had been observed.

Furthermore, NMFS, Marine Mammal Commission, and Scientific Review Group scientists expressed concern that strategic status decisions for some stocks may not be accurate because NMFS does not have all of the information needed to define the stocks accurately. Under the MMPA, marine mammal species are treated as stocks—populations located in a common area that interbreed when mature. However, a 2004 NMFS report found that the stock definitions for 61 percent of marine mammal stocks were potentially not accurate.[21] For example, a stock definition would not be accurate if NMFS defined two distinct populations of a marine mammal species incorrectly as one stock. If one of these two populations has a high level of incidental take and the other does not, the combined human-caused mortality estimate might not be high enough to result in a strategic status determination. However, if the two distinct populations were defined as two stocks, the high incidental take of one stock could result in it being considered strategic and triggering one of the requirements for take reduction team establishment. The Alaska Scientific Review Group has raised concerns that inaccurate stock definitions may be leading to incorrect strategic status designations. Specifically, in a 2007 letter to NMFS, the review group said that recent scientific information indicates that the current stock definitions might inappropriately consolidate harbor seal populations in Alaska. The review group chair said that this consolidation may lead to some harbor seal populations being incorrectly categorized as nonstrategic.

Outdated Information Reduces the Reliability of NMFS's Strategic Status Determinations

Our review of a sample of stock assessment reports found that approximately 11 of the 74 stocks used outdated information—information that is 8 years old or older—to calculate the maximum removal level, thereby reducing the reliability of the strategic status determinations for these stocks.[22] According to NMFS guidelines, information that is 8 years old or older is generally unreliable for estimating the current stock population. NMFS scientists estimate the size of a stock's population to determine its maximum removal level. If human-caused mortality exceeds maximum removal levels, the stock is considered strategic. However, when the data are 8 years old or older, scientific research has shown that marine mammal stocks could have declined significantly since the data were collected.[23] This could lead NMFS to inaccurately designate a stock as nonstrategic and therefore not establish a take reduction team when one might be needed. In addition, if a stock's population has increased significantly during the time period since the last estimate was made, NMFS may inaccurately designate the stock as strategic. Furthermore, our review found that for approximately 21 of the 74 stocks, the population size information was between 5 and 8 years old, a situation that is less of a concern than data that are 8 years old or older, but could also lead NMFS to make an inaccurate strategic stock determination. NMFS and Marine Mammal Commission scientists stated that scientists' confidence in the accuracy of the information used to estimate population size begins to decrease even before 8 years. Also, a 2004 NMFS report to Congress stated that estimates for population size based on information 5 years old or older may not accurately represent a marine mammal stock's current population size.[24]

Imprecise Information Reduces the Reliability of NMFS's Strategic Status Determinations

Our review of a sample of stock assessment reports from 2007 frequently found that NMFS used population size or fishery-related mortality estimates that were less precise than NMFS's guidelines recommend, decreasing the likelihood that strategic status determinations based on this information are accurate. Furthermore, we also found that NMFS could often not identify the level of precision for fishery-related mortality estimates. Specifically, we found that

- Approximately 48 of 74 stocks had population size estimates—used to determine maximum removal levels—that were less precise than NMFS guidelines recommend.[25] According to NMFS officials, one reason for the lack of precision is that the agency did not have adequate funding to conduct thorough population surveys. When conducting a marine mammal population survey, scientists document how frequently they observe marine mammals during a set period of time and use this information to estimate total population size. The duration of the survey and the number of scientists observing different areas within the stock's natural habitat affect the extent to which the survey is thorough and the population estimate is precise.

- Scientists could not calculate the precision of fishery-related mortality estimates—used to determine human-caused mortality estimates—for approximately 48 of the 74 stocks. In addition, the estimates for approximately 24 of the remaining 26 stocks were less precise than NMFS guidance recommends. Specifically, precision cannot be calculated when the sources of mortality data are anecdotal or the fishery-related mortality estimate is zero.[26] For these cases, NMFS does not have a systematic way of determining how precise the estimates are and therefore how much certainty it should have in their accuracy. NMFS and Marine Mammal Commission officials identified inadequate observer coverage as one of the main reasons for imprecise mortality estimates. According to National Observer Program officials, 52 percent of Category I or II fisheries have observer coverage; however, only 27 percent of Category I or II fisheries have adequate or near-adequate coverage levels.[27] Without adequate observer coverage in fisheries likely to cause incidental take of marine mammals, estimates will be less precise because they will be based on fewer data. NMFS and Marine Mammal Commission officials also stated that current funding levels for the observer program are inadequate to gather enough data on fishery-related mortality.

For the stocks for which we found that NMFS could calculate the level of precision for population size or fishery-related mortality estimates but these estimates were less precise than NMFS's guidance recommends, NMFS policy guidelines directed scientists to make adjustments to these estimates that increased the likelihood that the stocks were categorized as strategic. By doing this, the imprecision in these estimates is less likely to lead NMFS to overlook a stock that should be covered by a take reduction team, but NMFS officials told us that it is possible these stocks would not be designated as strategic if more precise estimates had been available and therefore these adjustments had not been necessary. However, in the approximately 48 of 74 stocks where NMFS cannot calculate the precision of a fishery-related mortality estimate—even though high levels of imprecision may exist—it cannot make these adjustments and therefore may either overlook some stocks that should be designated as strategic or inaccurately designate others as nonstrategic. Figure 2 summarizes key data limitations identified earlier in this chapter.

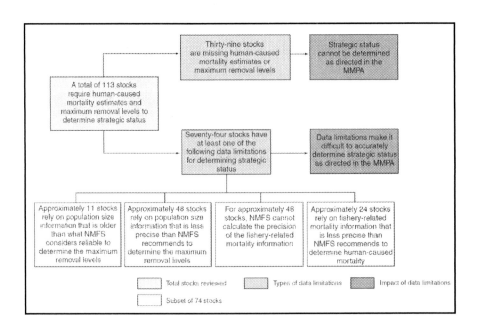

Source: GAO analysis of NMFS data.

Figure 2. Information Limitations Make It Difficult to Accurately Determine Strategic Status.

Funding Constraints Limit NMFS's Ability to Gather Sufficient Data, But the Agency Has Taken Some Steps to Identify Data Needs

NMFS officials acknowledged limitations in the information available to determine strategic status and the potential consequences, but identified funding constraints as an impediment to addressing these limitations. Specifically, a NMFS official stated that the agency has insufficient data to make informed management decisions regarding marine mammals in most instances, and said that a stock with sufficient data is an exception. However, while NMFS officials acknowledged these significant data limitations and their potential consequences, they also stated that they use the best scientific information available to make these determinations, as required by the MMPA. In addition, NMFS and Marine Mammal Commission officials stated that funding constraints have limited the agency's ability to gather the data that it needs to make accurate decisions about which stocks meet the statutory requirements for establishing take reduction teams.

NMFS has taken some steps to identify data limitations and proposed some actions to alleviate them. For example, a 2004 NMFS study found that the agency must significantly enhance the quantity and quality of its stock assessment data and analyses to meet MMPA mandates and outlined the actions and resource increases necessary to achieve these enhancements.[28] According to NMFS officials, the agency received funding to begin implementing the study's recommendations in fiscal year 2008 but the program lost other funding sources, so the new funding did not result in an overall increase in resources to improve these data. In addition, NMFS is currently completing a study to assess its sources of fishery-related mortality information. According to agency documents, this chapter will include an evaluation of the adequacy of the scientific techniques and existing observer coverage levels used to collect these data.

Nonetheless, marine mammal scientists expressed interest in having more information about the quality of the data used to determine the strategic status for each stock. Specifically, Marine Mammal Commission officials supported implementing a process to identify stocks where the scientists have low confidence in the quality of the data. According to these officials, if this occurred, interested parties would gain a better understanding of the data underlying strategic status determinations and therefore would have more information to judge the usefulness of the conclusions made from those data. Also, marine mammal scientists said that a process to identify stocks with poor data could make it easier to highlight stocks in need of additional data collection efforts.

NMFS HAS NOT ESTABLISHED TEAMS FOR MANY MARINE MAMMALS THAT HAVE MET THE STATUTORY REQUIREMENTS OF THE MMPA FOR A VARIETY OF REASONS

On the basis of NMFS's available information, we identified 30 marine mammal stocks that met the MMPA's requirements for establishing a team, and NMFS has established six teams that cover 16 of them. NMFS has not established teams for the 14 other marine mammals that have met the MMPA's requirements for establishing a team for several reasons: (1) the agency lacked sufficient funds to establish a team, (2) information about the stock's population size or mortality is outdated or incomplete and the agency lacks funds to obtain better information, (3) commercial fisheries account for little or no incidental take, or (4) the population size is increasing; therefore establishing a team for the stock is a lower priority.

Table 2. Marine Mammals That Have Met the MMPA's Requirements for Establishing a Take Reduction Team and for Which NMFS Has Established a Team

Take reduction team name	Marine mammals	Types of fisheries affected
Atlantic Large Whale	Fin whale Humpback whale North Atlantic right whale	Multiple trap/pot fisheries Multiple gillnet fisheries
Atlantic Trawl Gear	Short-finned pilot whale Long-finned pilot whale	Multiple trawl fisheries
Bottlenose Dolphin	Bottlenose dolphin	Multiple gillnet fisheries One trap/pot fishery Two seine fisheries

		Two stop/pound net fisheries
Harbor Porpoise[a]	Harbor porpoise	Multiple gillnet fisheries
Pacific Offshore Cetacean	Mesoplodont beaked whale Baird's beaked whale Cuvier's beaked whale Sperm whale Pygmy sperm whale Humpback whale Short-finned pilot whale Fin whale Long-beaked common dolphin	One gillnet fishery
Pelagic Longline	Short-finned pilot whale[b] Long-finned pilot whale[b]	The Atlantic portion of one longline fishery

Source: GAO analysis of *Federal Register* and NMFS documents.

[a] NMFS has established two take reductions teams for harbor porpoises. According to the *Federal Register*, the first team was established on February 12, 1996, to address harbor porpoises in the Gulf of Maine (known as the Gulf of Maine Harbor Porpoise take reduction team). The second team, established on February 25, 1997, focused on the same stock of harbor porpoises in the mid-Atlantic (known as the Mid-Atlantic take reduction team). NMFS decided to combine the two teams into one team in December 2007. For the purposes of this chapter, we refer to the combined teams as the Harbor Porpoise team.

[b] These marine mammal stocks are covered by both the Atlantic Trawl Gear and Pelagic Longline take reduction teams. Since we are reporting on the number of marine mammals that met the MMPA's requirements for establishing a team, we accordingly did not double-count this marine mammal.

NMFS Has Established Teams for 16 Marine Mammals That Have Met the Requirements of the MMPA

Since 1994, NMFS has established eight take reduction teams, six of which are still in existence—the Atlantic Large Whale, Atlantic Trawl Gear, Bottlenose Dolphin, Harbor Porpoise, Pacific Offshore Cetacean, and Pelagic Longline.[29] These six teams cover 16 marine mammal stocks that have met the MMPA's requirements for establishing a take reduction team. The MMPA gives NMFS discretion to determine how teams can be structured. For example, NMFS can establish a take reduction team for (1) one stock that interacts with multiple fisheries, such as the Bottlenose Dolphin take reduction team; (2) multiple stocks within a region, such as the Atlantic Large Whale take reduction team; or (3) multiple stocks that interact with one fishery, such as the Pacific Offshore Cetacean take reduction team. The existing take reduction teams—five of which are located in the Atlantic region and one in the Pacific—are described in table 2.

NMFS Has Not Established Teams for 14 Other Marine Mammals That Have Also Met the Requirements of the MMPA

NMFS has not established take reduction teams for 14 other marine mammals that have also met the MMPA's requirements for the establishment of a take reduction team. Table 3 lists these 14 marine mammals.

Table 3. Marine Mammals Stocks That Have Met the MMPA's Requirements for Establishing a Take Reduction Team, but NMFS Has Not Established a Team

Marine mammal	Stock
Bottlenose dolphin	Gulf of Mexico bay, sound, and estuarine
Bottlenose dolphin	Northern Gulf of Mexico coastal
Cuvier's beaked whale	Western North Atlantic
False killer whale	Hawaii
Harbor porpoise	Bering Sea
Harbor porpoise	Gulf of Alaska
Harbor porpoise	South East Alaska
Humpback whale	Central North Pacific
Humpback whale	Western North Pacific
Mesoplodont beaked whale	Western North Atlantic
Sperm whale	Hawaii
Northern fur seal	East North Pacific
Steller sea lion	Eastern United States
Steller sea lion	Western United States

Source: GAO analysis of *Federal Register* and NMFS data.

NMFS has not established teams for these 14 marine mammal stocks for the following reasons:

Lack of funding. Specifically, NMFS officials told us they did not establish a take reduction team for one marine mammal—the false killer whale—due to lack of funding. False killer whales found in the waters off the Hawaiian Islands have met the MMPA's requirements for establishing a team since 2004 because the stock has been strategic and interacts with a Category I longline fishery. Furthermore, since 2004, estimates of fishery-related mortality of false killer whales are at levels greater than their maximum removal level, according to stock assessment reports. According to the most recently available information, the false killer whale is the only marine mammal for which incidental take by commercial fisheries is known to be above its maximum removal level that is not covered by a take reduction team.[30] Since 2003, the Pacific Scientific Review Group has recommended that NMFS establish a team for these whales. Although NMFS officials told us that in accordance with the MMPA, the false killer whales are their highest priority for establishing a team, they said the agency does not have the funds to do so. NMFS officials told us the agency instead decided to focus what they characterized as their very limited funding on the already established take reduction teams. However, in the absence of a take reduction team, the Hawaii longline fishery continues to operate without a take reduction plan designed to reduce incidental take of false killer whales.

Outdated or incomplete data. NMFS has not established take reduction teams for eight marine mammals that interact with commercial fisheries in the Gulf of Mexico and the waters off of Alaska's coast because the information the agency has on them is too outdated or incomplete for agency officials to determine whether these marine mammals should be considered a high priority for establishing a take reduction team. Also, take reduction team members need better information about mortality before they can propose changes to fishing practices in a draft take reduction plan. However, because take reduction teams have not been established for these eight marine mammal stocks, fisheries continue to operate without take reduction plans that could decrease incidental take of these stocks.

Specifically, NMFS has not established teams for two stocks of bottlenose dolphins found in the Gulf of Mexico and six stocks in the waters off Alaska's coast, including three stocks of harbor porpoises, two stocks of Steller sea lions, and one stock of humpback whales.[31] Two stocks of bottlenose dolphins found in the Gulf of Mexico have met the MMPA's requirements for establishing a team since 2005 because they have been strategic and interact with two Category II fisheries.[32] According to stock assessment reports, the best scientific information available about population size for these two stocks is 8 years old or older. According to NMFS documents, using such outdated information increases the possibility that significant population changes of which NMFS is unaware could have occurred. Agency officials told us that the 2008 survey to collect new population size estimates was canceled due to insufficient funding. Furthermore, according to stock assessment reports, the available mortality estimates are incomplete because they are based on anecdotal information. Consequently, scientists can use this information only to make a minimum estimate of the number of marine mammals being killed or injured. Agency officials told us they would like to begin observing the two Gulf of Mexico fisheries, but are currently unable to do so due to funding constraints.

Similarly, NMFS has not established take reduction teams due to outdated information for three stocks of harbor porpoises found in the waters off Alaska's coast that have met the MMPA's statutory requirements for establishing a team since 2006 because they have been strategic and interact with multiple Category II fisheries. According to stock assessment reports, the best scientific information available about population size for harbor porpoises is outdated because the estimates are 8 years old or older. NMFS officials told us harbor porpoises are a major conservation concern for the agency, but they said funding constraints have limited their ability to collect new population size estimates for these marine mammals.

In addition, NMFS has not established take reduction teams due to incomplete information for two stocks of Steller sea lions that have met the MMPA's requirements for establishing a team since 1996 because they have been strategic and interact with multiple Category II fisheries. NMFS officials told us the fishery-related mortality information for these stocks is incomplete because they are uncertain whether incidental take is occurring in commercial fisheries not covered by observer programs. According to these same officials, lack of funding has limited the agency from obtaining more complete fishery-related mortality information for Steller sea lions.

Last, NMFS has not established a take reduction team due to outdated information for the Western North Pacific stock of humpback whales, which has met the MMPA's requirements for establishing a team since 2006, because it has been strategic and interacts with two Category II fisheries. According to the stock assessment report, the best scientific information available about population size for these humpback whales is outdated because it is 8 years old or older, but agency officials told us funding constraints limit their ability to collect new information.

Commercial fisheries account for little or no incidental take. NMFS has not established teams for four marine mammals—the Hawaii stock of sperm whales, Western North Atlantic stocks of Cuvier's beaked whales and Mesoplodont beaked whales, and East North Pacific stock of northern fur seals—that have met the MMPA requirements for establishing a team because, according to agency officials, commercial fisheries account for little or no incidental take of these stocks. According to our analysis, these sperm whales meet the statutory requirements for a team because they are listed as an endangered species under the ESA, and therefore are a strategic stock, and they interact with a Category I fishery. However, NMFS officials told us that the commercial fishery with which these sperm whales interact accounts for little or no incidental take, and therefore it would be inappropriate to establish a team for them.[33] Similarly, NMFS's 2007 stock assessment reports state that acoustic activities, such as sonar used by the U.S. Navy, may contribute to the mortality and serious injury of Cuvier's and Mesoplodont beaked whales, and non-human-related causes of death that are unknown to scientists are contributing to the population decline of northern fur seals. NMFS officials told us it would be inappropriate to establish take reduction teams for these marine mammal stocks because mortality and serious injuries are not being caused by interaction with a commercial fishery. According to NMFS officials, they proposed amending the MMPA in 2005 to require that take reduction teams be established only for strategic stocks that interact with Category I or II fisheries and that have some level of fishery-related incidental take of those stocks, but Congress took no action on the proposal at that time.

Population size is increasing. NMFS officials said they have not established a take reduction team for one marine mammal stock that meets the statutory requirements—the Central North Pacific stock of humpback whales––because of insufficient funding; however, this stock would be a low priority because the stock's population size is increasing. This stock is strategic because it is listed as an endangered species under the ESA and it interacts with a Category I fishery off the coast of Hawaii and multiple Category II commercial fisheries in the waters off Alaska's coast. However, because its population size is increasing, NMFS officials consider the stock to be a lower priority for establishing a team than stocks with declining populations.

Table 4. Delays in Establishing Take Reduction Teams

Take reduction team	Date of statutory deadline for team establishment	Date take reduction team was established	Delay
Atlantic Large Whale	May 1, 1996[a]	August 6, 1996	97 days (3 months)
Pelagic Longline	January 2002[b]	June 22, 2005[c]	1,268 days (over 3 years)

Bottlenose Dolphin	May 1, 1996[a]	October 24, 2001[d]	2,001 days (over 5 years)

Source: GAO analysis of information published in *Federal Register* notices.

[a] The stocks covered by this team were designated as strategic in the July 1995 stock assessment reports. The effective date of the list of fisheries that identified these stocks as interacting with Category I or II fisheries was April 1, 1996. Accordingly, the deadline for team establishment is 30 days after the list of fisheries' effective date.

[b] The stock assessment reports for the relevant marine mammal stocks covered by the Pelagic Longline were published in December 2001. However the reports did not include a specific date in December. The stocks covered by this team were listed as strategic in the December 2001 stock assessment reports. Also, the list of fisheries that was in effect at that time listed the stocks as interacting with Category I or II fisheries. Accordingly, the deadline for team establishment is 30 days after publication of the stock assessment reports. We determined the number of days of delay based on the date January 1, 2002, because the December 2001 stock assessment reports did not indicate a specific date.

[c] This team was established pursuant to an agreement settling a lawsuit.

[d] The Bottlenose Dolphin team was originally established on August 31, 2001; however, due to the events of September 11, 2001, the first meeting was canceled and NMFS subsequently reestablished the team on October 24, 2001.

NMFS HAS HAD LIMITED SUCCESS IN MEETING THE STATUTORY DEADLINES FOR TAKE REDUCTION TEAMS FOR A VARIETY OF REASONS

For the five take reduction teams subject to the MMPA's deadlines, NMFS has had limited success in meeting the deadlines for a variety of reasons.[34] NMFS did not meet the statutory deadlines to establish take reduction teams for three of the five teams, in one case due to a lack of information about population size or mortality. In addition, two of the five teams did not submit their draft take reduction plans to NMFS within the statutory deadlines, in one case because the team could not reach consensus on a plan. NMFS did not publish proposed take reduction plans within the statutory deadlines for any of the five teams because of the time needed to complete the federal rulemaking process, among other things. However, NMFS has complied with the statutory deadlines for the public comment periods on the proposed plans for all five teams. Finally, NMFS did not publish final take reduction plans within the statutory deadlines for four of the five teams because of the time associated with analyzing public comments, among other things.

NMFS Did Not Establish Three of the Five Teams Within the Statutory Deadlines

According to the MMPA, NMFS has 30 days to establish a take reduction team after a stock is listed as strategic in a final stock assessment report and is listed as interacting with a Category I or II fishery in the current list of fisheries. NMFS established two teams within this statutory deadline: the Harbor Porpoise and Pacific Offshore Cetacean. However, NMFS did not meet the statutory deadlines for establishing three teams—the Atlantic Large Whale, Pelagic Longline, and Bottlenose Dolphin. These teams were established from 3 months to more than 5 years after their statutory deadlines (see table 4).

According to NMFS officials, the reasons for delays in establishing these take reduction teams include the following:

- **Atlantic Large Whale:** It took NMFS officials more than 30 days to identify sufficient take reduction team members to represent the stocks' large habitat, which stretches from Maine to Florida.
- **Pelagic Longline:** After 2001, NMFS officials were waiting to see if modifications to the longline fishery, intended to reduce the incidental take of billfish and sea turtles, would also reduce incidental take of pilot whales, which would eliminate the need for this team.[35] However, in 2002, an environmental group sued NMFS because of the agency's alleged failure to establish take reduction teams for marine mammals that met the statutory requirements. According to an agreement settling the lawsuit, NMFS had to conduct surveys and develop new population size estimates for pilot whales. In addition, it had to establish a take reduction team for the Atlantic portion of a large pelagic longline fishery by June 30, 2005.

- **Bottlenose Dolphin:** NMFS lacked information about population size and mortality for bottlenose dolphins that take reduction team members need to consider before they can propose changes to fishing practices in a draft take reduction plan, and NMFS scientists recommended that the agency obtain better information before establishing a team.[36] According to a NMFS official, mortality information for bottlenose dolphins collected between 1995 and 1998 was published in the 2000 stock assessment report. As a result of this new information, NMFS established a team in 2001.

Two Teams Did Not Develop and Submit Draft Take Reduction Plans within the Statutory Deadlines

According to the MMPA, after NMFS establishes a take reduction team, the team must develop a draft take reduction plan and submit it to NMFS within 6 months if it covers strategic stocks whose level of human-caused mortality exceeds the maximum removal level. However, if the level of human-caused mortality for strategic stocks covered by the plan is below the maximum removal level, as it is for the Pelagic Longline team, then the team has 11 months to develop a draft plan and submit the draft plan to NMFS. Three of the five teams submitted their draft plans within the statutory deadlines.[37] However, two teams—the Pelagic Longline and Bottlenose Dolphin—submitted their draft take reduction plans to NMFS, 17 and 23 days respectively, after their statutory deadlines (see table 5). Table 5 shows the delays in developing and submitting draft plans for the two take reduction teams that missed the statutory deadline.

According to NMFS officials, the reasons for delays in submitting draft take reduction plans to NMFS include the following:

- **Pelagic Longline:** The unexpected death of a take reduction team member 1 week before the plan's due date delayed the team's submission to NMFS. This team member was a key liaison to the fishing industry, working with commercial fishermen to obtain agreement on potential take reduction plan measures.

- **Bottlenose Dolphin:** The take reduction team found it difficult to reach consensus about modifications to fishing practices to help reduce incidental take because of the large number of team members involved (44) representing multiple types of fisheries. For example, the Bottlenose Dolphin team includes four gillnet, one trap/pot, two seine, and two stop/pound net fisheries, making it difficult to agree on modifications to fishing practices. See appendix II for a description of these fishing techniques.

NMFS Did Not Publish Proposed Take Reduction Plans for Five Teams within the Statutory Deadlines for a Variety of Reasons

According to the MMPA, once NMFS receives a draft take reduction plan, it must publish a proposed plan and implementing regulations in the Federal Register within 60 days. NMFS missed the statutory deadline for publishing proposed plans and implementing regulations for all five teams by 5 days to more than 2 years after the statutory deadlines (see table 6).

Table 5. Delays in Developing and Submitting Draft Take Reduction Plans

Take reduction team	Date of statutory deadline for submission of draft plan	Date draft plan submitted to NMFS	Delay
Pelagic Longline	May 22, 2006	June 8, 2006	17 days
Bottlenose Dolphin	April 24, 2002	May 17, 2002[a]	23 days

Source: GAO analysis of information published in *Federal Register* proposed rules.

[a] The *Federal Register* notice states that the draft plan was submitted on May 17, 2002; NMFS's records indicate that the team's facilitator submitted a draft plan to NMFS on May 6, 2002; however, the date on the draft plan is May 7, 2002. Due to discrepancies in the various records, we relied on the date in the *Federal Register* notice to determine the deadlines. An addendum to the plan was submitted to NMFS on May 3, 2003, after the team reconvened on April 1-3, 2003, to discuss new scientific information. We used the date of the original submission because NMFS was not obligated to reconvene the team to address the new information. Under the MMPA, NMFS has the statutory authority to issue a proposed plan that departs from a team's draft plan.

Table 6. Delays in Publishing Proposed Take Reduction Plans

Take reduction team	Date of statutory deadline for publication[a]	Date NMFS published the proposed plan in Federal Register	Delay
Atlantic Large Whale	April 2, 1997	April 7, 1997	5 days
Pacific Offshore Cetacean	October 14, 1996	February 14, 1997	123 days (4 months)
Harbor Porpoise	March 15, 1998[b]	September 11, 1998	180 days (6 months)
Pelagic Longline	August 7, 2006	June 24, 2008	686 days (almost 2 years)

Bottlenose Dolphin	July 16, 2002	November 10, 2004	847 days (over 2 years)

Source: GAO analysis of information published in *Federal Register* proposed rules.

[a] This deadline was calculated based on 60 days after the team members submitted a draft plan to NMFS. Two teams, the Pelagic Longline and Bottlenose Dolphin, submitted their draft plans to NMFS late, but we calculated this deadline based on 60 days after the team submitted a draft plan to NMFS, not based on 60 days after the prescribed deadline.

[b] NMFS established the Gulf of Maine take reduction team on February 12, 1996, and the Mid-Atlantic take reduction team on February 25, 1997. The current Harbor Porpoise take reduction team is a combination of these two prior teams that focused on harbor porpoises in distinct geographic areas, the Gulf of Maine and the mid-Atlantic. The original draft plan for the Gulf of Maine take reduction team was submitted to NMFS on August 8, 1996. Then the Mid-Atlantic team submitted its draft plan to NMFS on August 25, 1997. The Gulf of Maine take reduction team developed and submitted a second draft take reduction plan on January 14, 1998. The Mid-Atlantic take reduction team recommendations were then incorporated into this January 14, 1998, draft plan as a combined plan. We consider this last date, January 14, 1998, as the submission date for a draft plan because at that point both teams had concluded their deliberations.

According to NMFS officials, the reasons for delays in publishing proposed plans and implementing regulations include the following:

- **Atlantic Large Whale:** Agency officials submitted the proposed plan for publication within the statutory deadline but told us that the *Federal Register* did not print the notice containing the proposed take reduction plan until 5 days after the statutory deadline.

- **Pacific Offshore Cetacean:** The former team coordinator for this team said that the proposed plan was delayed because of the time it took to comply with various applicable laws. For example, NMFS is required to review the proposed plan and consider its effects on small businesses and other small entities under the Regulatory Flexibility Act and prepare an environmental assessment under the National Environmental Policy Act, among other requirements. Developing and drafting an environmental assessment is a labor-intensive process requiring coordination among scientists, economists, and policymakers.

- **Harbor Porpoise:** According to NMFS officials, they delayed preparing the proposed plan for publication in the *Federal Register* because they were waiting to see if closures of some fishing areas to protect fish would also reduce incidental take of harbor porpoises. In addition, NMFS scientists determined that this stock of harbor porpoises was migratory and interacting not only with the Gulf of Maine fisheries but with mid-Atlantic fisheries as well. As a result of this finding, NMFS established another team, the Mid-Atlantic take reduction team, for the mid-Atlantic fisheries. NMFS delayed the publication of the proposed take reduction plan for the Gulf of Maine fisheries until the Mid-Atlantic team developed and submitted a draft plan. Ultimately, the two plans were combined and published as a single plan for both the Gulf of Maine and mid-Atlantic fisheries.

- **Pelagic Longline:** According to NMFS officials, a combination of factors caused the proposed plan to be published in the *Federal Register* almost 2 years after the deadline. Take reduction team coordinators are responsible for coordinating NMFS's internal review and approval for take reduction plans, crafting the regulatory language for the plan, and submitting the proposed plans for publication in the *Federal Register*. Because the team coordinator position was vacant for approximately 16 months, completion of these tasks was delayed.

- **Bottlenose Dolphin:** A combination of factors caused this proposed plan to be published in the *Federal Register* 2 years after the deadline, according to NMFS officials. The publication of the proposed plan was delayed because NMFS asked team members to reconvene when it became clear that the recommended regulatory measures would not reduce incidental take to levels below the maximum removal level, as required by the MMPA. Although NMFS can propose a plan of its own that deviates from the team's draft plan, officials from NOAA's Office of General Counsel told us NMFS prefers to wait until the team completes its work and submits a draft plan. After they reconvened, the take reduction team members developed and submitted a revised draft plan; however, because the team coordinator position was vacant for about 8 months, preparing the proposed plan for publication was delayed. Additionally, because NMFS combined two rules—to benefit both sea turtles and bottlenose dolphins—into one, the proposed plan was delayed due to the time needed to update an environmental assessment required under the National Environmental Policy Act and other associated documents. Also, the proposed plan was delayed because of the time it took to comply with various laws and executive orders. Finally, the Office of Management and Budget took 90 days to review the proposed plan—the maximum time allowed for such a review. This review by itself exceeded the MMPA's 60-day deadline.

NMFS officials told us it is extremely difficult for the agency to meet the MMPA's deadline for this step in the process. As the examples above demonstrate, NMFS officials provided us with a variety of reasons for delays in meeting the statutory deadlines for publishing proposed plans in the *Federal Register*; however, the agency has not conducted a comprehensive analysis that would identify all of the tasks that must be completed during this stage in the process, along with the total time needed to complete them. NMFS stated that it has not conducted such an analysis because, in some cases, the documents needed are 10 years old and are not available electronically.

NMFS Has Complied with the Statutory Deadlines for Public Comment Periods

According to the MMPA, NMFS must hold a public notice and comment period on the proposed plan and implementing regulations for up to 90 days after the proposed plan's publication in the *Federal Register*. The public comment period is an opportunity for interested persons to participate in the development of a take reduction plan by submitting their views and concerns about the proposed plan. For all five teams—the Atlantic Large Whale, Bottlenose Dolphin, Harbor Porpoise, Pacific Offshore Cetacean, and Pelagic Longline—NMFS has complied with the statutory deadline each time.

NMFS Did Not Publish Final Take Reduction Plans for Four of the Five Teams within the Statutory Deadlines

According to the MMPA, once the public comment period ends, NMFS must publish the final plan and implementing regulations in the *Federal Register* within 60 days. NMFS missed the statutory deadline for four teams but met it for the Harbor Porpoise team. According to our analysis, the delays ranged from 8 days to just over 1 year (see table 7).

Table 7. Delays in Publishing Final Take Reduction Plans

Take reduction team	Date of statutory deadline for publication	Date NMFS published the final plan in Federal Register	Delay
Atlantic Large Whale	July 14, 1997	July 22, 1997[a]	8 days
Pacific Offshore Cetacean	May 30, 1997	October 3, 1997	136 days (4 months)
Bottlenose Dolphin	April 9, 2005	April 26, 2006	382 days (over 1 year)
Pelagic Longline	November 21, 2008	To be determined[b]	To be determined

Source: GAO analysis of information published in *Federal Register* final rules.

[a] NMFS published an "interim final plan" for the Atlantic Large Whale team. Although the MMPA is silent on interim final plans, we consider it the final plan because it was in force and had the same effect as a final plan.

[b] NMFS had not published the final plan in the *Federal Register* as of the publication date of our report, December 8, 2008.

According to NMFS officials, the reasons for delays in publishing final plans and implementing regulations in the *Federal Register* include the following:[38]

- **Atlantic Large Whale:** The delay was due, in part, to NMFS's efforts in responding to the large number of public comments received on the proposed plan.

- **Pacific Offshore Cetacean:** Because the plan included a fishing gear modification, NMFS waited until the preliminary results of a gear research experiment indicated that the modification reduced incidental take before publishing the final plan. The experiment tested the effectiveness of acoustic devices, known as pingers, that are attached to fishing nets and emit high-pitched sounds so that marine mammals would avoid the nets.

- **Bottlenose Dolphin:** According to NMFS officials, the delay was the result of the time needed to review and analyze over 4,000 comments the agency received during the public comment period and the 90 days the Office of Management and Budget took to review the final take reduction plan before NMFS could publish it in the *Federal Register*.

NMFS Does Not Have a Comprehensive Strategy for Evaluating the Effectiveness of Take Reduction Regulations

NMFS does not have a comprehensive strategy—identified as a key principle by the Government Performance and Results Act of 1993—for assessing the effectiveness of take reduction regulations once they have been implemented. The Government Performance and Results Act of 1993 provides a foundation for examining agency performance goals and results. Our work related to the act and the experience of leading organizations have shown the importance of developing a comprehensive strategy for assessing program effectiveness that includes, among other things, program performance goals that identify the desired results of program activities and reliable information that can be used to assess results.[39] In the context of NMFS's efforts to measure the success of take reduction plans and implementing regulations, such a strategy would include, at a minimum, (1) performance goals that identify the desired outcomes of the take reduction regulations; (2) steps for assessing the effectiveness of potential take reduction regulations, such as fishing gear modifications, in achieving the goals; (3) a process for monitoring the fishing industry's compliance with the requirements of the take reduction regulations; and (4) reliable data assessing the regulation's effect on achieving the goals. Instead of such a comprehensive strategy, we found that although NMFS uses short- and long-term goals established by the MMPA to evaluate the success of take reduction regulations, these goals and the data that NMFS uses to measure the impact of the take reduction regulations have a number of limitations. In addition, while NMFS has taken steps to identify the impact of proposed take reduction regulations prior to their implementation, the agency has limited information about the fishing industry's compliance with the regulations once they have been implemented. As a result, when incidental takes occur in fisheries covered by take reduction regulations, it is difficult for NMFS to determine whether the regulations were not effective in meeting the MMPA's goals or whether the fisheries were not complying with the regulations.

Limitations in the Goals and Data That NMFS Currently Uses to Evaluate the Success of Take Reduction Regulations Impede Effective Program Evaluations

The MMPA identifies, and NMFS further defines, short- and long-term goals for reducing incidental take of marine mammals that take reduction regulations should achieve. Specifically, the MMPA set a short-term goal of reducing incidental take—also known as fishery-related mortality—for strategic stocks below the maximum removal level within 6 months of a plan's implementation and set a long-term goal of reducing fishery-related mortality to insignificant levels approaching a zero mortality and serious injury rate within 5 years of a plan's implementation, which NMFS generally defines as 10 percent of the maximum removal level.[40] NMFS officials told us that NMFS staff and take reduction team members review whether or not the goals have been met for the stocks covered by their teams.

However, NMFS officials, Marine Mammal Commission officials, and a Scientific Review Group chair all considered the 6-month time frame for meeting the short-term goal to be unrealistic. Specifically, some noted that due to the extensive time needed to gather and publish data on the maximum removal level and fishery-related mortality estimates, NMFS does not have the necessary information to assess the goal within the 6-month time frame. A NMFS official also noted that fishing changes over the year; therefore, assessing whether fishery-related morality is below the maximum removal level during a 6-month time frame may not consider mortality that may occur during both the busiest and the slowest fishing seasons. While the MMPA sets this 6-month goal, it does not impose consequences on NMFS or the regulated fisheries if the goal is not met.

Furthermore, these goals may not help NMFS assess the success of the regulations because we found that there was not always greater success in meeting the goals after take reduction regulations were implemented than before they were implemented. Also, if the goals had been met for a stock in a given year, in some cases the goals did not continue to be met in the following years. Specifically, we found that for two stocks,[41] the short-term goal had been met prior to the regulations being implemented but was no longer being met in 2007.[42] In addition, for two other stocks, the long-term goal had been met prior to implementation of the regulations,[43] but was no longer being met in 2007.[44] Furthermore, for two stocks, the short-term goal had been met and for two stocks, the long-term goal had been met in 2007, but these goals had already been met prior to implementation of the take reduction regulations.[45]

In cases where the goals were met prior to the implementation of take reduction regulations, the goals cannot be used to determine the regulations' impact on reducing take.

In addition, according to NMFS officials, changes to the marine environment that happen during the same time period as the implementation of take reduction regulations make it difficult to assess whether the regulations are the reason that the short- and long-term goals for a stock have been achieved or whether it was other changes. Specifically, state or federal fishing regulations unrelated to the take reduction team process may result in less fishing in the fisheries covered by the take reduction team. In such a scenario, fishery-related mortality may decrease because there are fewer opportunities for fishing vessels to interact with marine mammals. Therefore, a lower level of fishery-related mortality may lead to achievement of the MMPA's goals for a stock even if the take reduction regulations themselves were not the primary reason for the lower level of incidental take.

Moreover, limitations in some of the data used to determine whether the MMPA's short- and long-term goals for reducing incidental take by commercial fisheries have been met may lead to inaccurate conclusions about the effectiveness of the take reduction regulations.[46] We reviewed the stock assessment reports for 9 of the 10 strategic stocks and all 3 of the nonstrategic stocks covered by take reduction regulations and found that for 2007, the short-term goal for 4 of the 9 strategic stocks had been achieved and the long-term goal had been achieved for 3 of the 12 strategic and nonstrategic stocks.[47] However, we also found that the information used to determine the maximum removal level or the fishery-related mortality estimate for 6 of the 9 strategic stocks covered by these regulations was less precise than NMFS guidelines recommend. Because these are the two key sources of information for determining whether the MMPA's short- and long-term goals have been met, this imprecision may cause NMFS to incorrectly assess whether the take reduction regulations have been successful.

NMFS officials stated that limitations in the data make it difficult to know the reason for changes in meeting the goals from one year to another. For example, we found that the short-term goal for the Gulf of Maine stock of humpback whales covered by the Atlantic Large Whale take reduction team had been met prior to implementation of the take reduction regulations; however, according to the stock assessment report, it did not meet the goal in 2007.[48] Meanwhile, between the year prior to when the regulations were implemented and 2007, NMFS altered its stock definition for these marine mammals in a way that decreased the number of animals included in the population size estimate—a key aspect of determining the maximum removal level. This change made the maximum removal level much lower than it had been before the regulations were implemented, making it more difficult to achieve the goals. Because of this change in NMFS's approach to calculating the maximum removal level, it is difficult to determine whether ineffectiveness of the take reduction regulations or the change in the maximum removal level led to the short-term goal no longer being met for this stock.

NMFS Studies the Impact of Proposed Take Reduction Regulations prior to Their Implementation, but Has Limited Information about Industry Compliance

NMFS has assessed the likelihood that proposed take reduction regulations would achieve the short- and long-term goals of reducing incidental take for all four teams with final take reduction plans and regulations. Specifically, for all four plans, scientists evaluated whether key proposed measures for modifying fishing gear or changing the times or areas where fishing could occur were likely to decrease incidental take. For example, NMFS scientists analyzed data from previous incidental take in the gillnet fisheries of concern for bottlenose dolphins and found that incidental take had occurred at a higher rate on the vessels that used nets with larger mesh openings. Because this type of gear would be restricted under the proposed regulations, NMFS had reason to believe that these gear restrictions would result in reduced incidental take of bottlenose dolphins.[49] Similarly, according to the environmental assessment report for the Harbor Porpoise take reduction team, a controlled experiment tested the effectiveness of acoustic devices—often called pingers—attached to fishing nets. Pingers emit a high-pitched sound that harbor porpoises can hear, which results in them avoiding fishing nets. This experiment allowed NMFS scientists to predict that proposed regulations to implement pingers would likely result in a decline of incidental take.[50]

Although NMFS has conducted some assessments of the likelihood that proposed take reduction regulations will achieve the goals of reducing incidental take, they have limited information about the extent to which fisheries comply with take reduction regulations once they have been implemented. As a result, when incidental takes occur in fisheries covered by take reduction regulations, it is difficult for NMFS to determine whether the regulations were not effective in meeting the MMPA's goals or whether the fisheries were not complying with the regulations. Specifically, we determined that NMFS does not have comprehensive approaches to measure the extent to which fisheries comply with the regulations for the four take reduction plans that it has implemented. However, for all of these implemented regulations, NMFS has some—albeit limited—information from fisheries observer or enforcement programs that provide an indication of whether fisheries are complying with the regulations.[51] For example, when incidental take of harbor porpoises in the fisheries covered by the Harbor Porpoise take reduction team recently increased, NMFS scientists used observer information about incidental take to determine whether or not these takes occurred when vessels were complying with the requirement to use pingers on their nets.[52] However, according to the scientists, the usefulness of this information in determining actual compliance was limited because observers could only identify whether the pingers were attached to the net, not whether these pingers functioned properly. On the Pacific Offshore Cetacean team, the team coordinator stated that in the past, NMFS has received information from the observer program about fishing vessels monitored by observers that were not in compliance with the take reduction regulations. However, she stated that NMFS does not routinely review the observer information to identify or document the extent of these instances or estimate the extent of overall compliance with the take reduction regulations.

In addition to the information that it receives from the observer programs, NMFS receives some information about compliance from NOAA's Office of Law Enforcement, the U.S. Coast Guard, or state enforcement agencies. Specifically, team coordinators told us that officials from the U.S. Coast Guard attend take reduction team meetings to discuss instances where the agencies found vessels out of compliance with take reduction regulations during the course of their enforcement work. However this information is not generally extensive enough to provide overall assessments of the extent to which fisheries are complying with the regulations.

In 2007, we reported that NMFS lacked a strategy for assessing industry compliance with the Atlantic Large Whale team's take reduction plan, and we recommended that it develop one.[53] In response to our report, the team is beginning to develop a comprehensive approach to monitoring compliance. NMFS staff members are currently developing a plan for take reduction team members to review during their next meeting, which is planned for early 2009. No other take reduction teams are developing comprehensive approaches for monitoring compliance at this time.

CONCLUSIONS

NMFS faces a very large, complex, and difficult task in trying to protect marine mammals from incidental mortality and serious injury during the course of commercial fishing operations, as the MMPA requires. Without comprehensive, timely, and accurate population and mortality data for the 156 marine mammal stocks that NMFS is charged with protecting, NMFS may be unable to accurately identify stocks that meet the legal requirements for establishing take reduction teams, thereby depriving them of the protection they need to help recover or maintain healthy populations. Conversely, unreliable data may lead NMFS to erroneously establish teams for stocks that do not need them, wasting NMFS's limited resources.

For those stocks that meet the requirements for establishing take reduction teams, it is important that NMFS adhere to the deadlines in the MMPA, as delays in establishing teams and developing and finalizing take reduction plans could result in continued harm to already dwindling marine mammal populations. However, we recognize that it may not be useful to establish take reduction teams for those stocks that meet the MMPA requirements but whose population declines are not being caused by commercial fisheries. We also acknowledge it may not be possible for NMFS to meet some of the MMPA's deadlines given the requirements of other laws that NMFS must comply with in developing take reduction plans and the need for various levels of review and approval. Nonetheless, the MMPA's deadlines are clear, and unless the law is amended to address these situations, NMFS has a legal obligation to comply with them.

Finally, for stocks for which NMFS has developed take reduction regulations, it is important for NMFS to assess their effectiveness in reducing serious injury and mortality to acceptable levels. Doing so will require more comprehensive information about the fishing industry's compliance with take reduction regulations so that if the short- and long-term goals are not met, NMFS knows whether to attribute the failure to a flaw in the regulations or to noncompliance with them. Without a comprehensive strategy for assessing the effectiveness of its take reduction plans and implementing regulations and industry's compliance with them, NMFS may be missing opportunities to better protect marine mammals.

MATTERS FOR CONGRESSIONAL CONSIDERATION

To facilitate the oversight of NMFS's progress and capacity to meet the statutory requirements for take reduction teams, Congress may wish to consider taking the following three actions:

- direct the Assistant Administrator of NMFS to report on major data, resource, or other limitations that make it difficult for NMFS to accurately determine which marine mammals meet the statutory requirements for establishing take reduction teams; establish teams for stocks that meet these requirements; and meet the statutory deadlines for take reduction teams;
- amend the statutory requirements for establishing a take reduction team to stipulate that not only must a marine mammal stock be strategic and interacting with a Category I or II fishery, but that the fishery with which the marine mammal stock interacts causes at least occasional incidental mortality or serious injury of that particular marine mammal stock; and
- amend the MMPA to ensure that its deadlines give NMFS adequate time to publish proposed and final take reduction plans and implementing regulations while meeting all the requirements of the federal rulemaking process.

RECOMMENDATION FOR EXECUTIVE ACTION

We recommend that the Assistant Administrator of NMFS develop a comprehensive strategy for assessing the effectiveness of each take reduction plan and implementing regulations, including, among other things, establishing appropriate goals and steps for comprehensively monitoring and analyzing rates of compliance with take reduction measures.

AGENCY COMMENTS AND OUR EVALUATION

We provided a draft copy of this chapter to the Secretary of Commerce for review and comment. In response to our request, we received general, technical, and editorial comments from NOAA by email, which stated that the agency agreed with our recommendation that NMFS should develop a comprehensive strategy for assessing the effectiveness of each take reduction plan and the implementing regulations. We have incorporated the technical and editorial comments provided by the agency, as appropriate.

As we agreed with your office, unless you publicly announce the contents of this chapter earlier, we plan no further distribution until 30 days from the report date. At that time, we will send copies to the Secretary of Commerce, the Administrator of NOAA, and appropriate congressional committees, and other interested parties.

Sincerely yours,

Anu K. Mittal

Anu K. Mittal
Director, Natural Resources and Environment

APPENDIX I. OBJECTIVES, SCOPE, AND METHODOLOGY

The objectives of this review were to determine the extent to which (1) available data allow the National Marine Fisheries Service (NMFS) to accurately identify the marine mammal stocks that meet the Marine Mammal Protection Act's (MMPA) requirements for establishing take reduction teams, (2) NMFS has established take reduction teams for those marine mammal stocks that meet the statutory requirements, (3) NMFS has met the statutory deadlines established in the MMPA for the take reduction teams subject to the deadlines and the reasons for any delays, and (4) NMFS has developed a comprehensive strategy for evaluating the effectiveness of the take reduction plans that have been implemented.

To determine the extent to which available data allowed NMFS to accurately identify marine mammal stocks that meet the MMPA's requirements for establishing take reduction teams, we identified stocks for which NMFS lacked data on either the human-caused mortality and serious injury estimate (hereafter referred to as human-caused mortality estimate) or the potential biological removal levels (hereafter referred to as maximum removal levels).[54] To do this, we first reviewed all 156 stocks identified in NMFS's 2007 stock assessment reports and removed 19 stocks currently covered by take reduction teams. Then we removed 24 stocks that are listed as endangered or threatened under the Endangered Species Act (ESA) or designated as depleted under the MMPA because NMFS does not use information about human-caused mortality and the maximum removal level to make strategic status decisions for these stocks. We then reviewed the remaining 113 stocks to identify those that lacked either a human-caused mortality estimate or a maximum removal level. After identifying those that lacked human-caused mortality or maximum removal levels, we reviewed a sample of the remaining 74 stocks that did have these determinations to assess the reliability of the information used to determine human-caused mortality estimates and maximum removal levels.

We identified several key data elements in NMFS's stock assessment reports that the agency uses to determine human-caused mortality estimates and maximum removal levels:

- abundance estimates (population size estimates) and NMFS calculation of the estimates' precision,
- the age of data used to calculate population size estimates,
- fishery-related mortality and serious injury estimates (hereafter known as fishery-related mortality estimates) and NMFS calculation of the estimates' precision,
- adjustments made to the maximum removal level in order to account for fishery-related mortality estimate imprecision,
- information sources such as observer data used to calculate fishery-related mortality estimates, and
- qualitative information identified in the stock assessment reports about scientists' concerns regarding data strengths or limitations.

We also identified criteria for assessing the quality of these data elements using information from the MMPA and publications such as NMFS's guidelines for preparing stock assessment reports and stock assessment improvement plan and confirmed the criteria with NMFS officials.[55] While scientists and publications also identified bias in population size and mortality estimates as a potential data reliability problem, we did not assess the extent to which existing data sources included bias because data and accompanying criteria to make such an assessment were not available.

Table 8. Confidence Intervals for Estimates Based on GAO's Review of Selected Stock Assessment Reports

Characteristics	Estimated population total with this characteristic	95 percent confidence interval of the total estimate
Population estimates used information that was 8 years old or older	11	5 - 18
Population size estimates used information that was between 5 and 8 years old	21	12 - 29
Population size estimates were less precise than NMFS guidelines recommend	48	41 - 56
Scientists could not calculate the precision of fishery-related mortality estimates	48	38 - 58
Mortality estimates were less precise than NMFS guidance recommends	24	14 - 34

Source: GAO analysis.

We then analyzed the key data elements for a sample of stocks to determine the extent to which the data met the criteria we identified. We chose our sample of stocks to review by conducting a stratified random sample of the 74 stocks that were not currently covered by take reduction teams, did not receive strategic status due to MMPA designations or listings under the ESA, and had both human-caused mortality and serious injury estimates and maximum removal levels. The sample of 28 stocks included all strategic stocks that met these criteria as well as a representative sample of stocks from each of the three NMFS Fishery Science Centers responsible for publishing the stock assessment reports. We then extrapolated the results of our review for this sample to all 74 stocks that met the criteria listed above. We calculated 95 percent confidence intervals for each of the estimates made from this sample. The confidence intervals for these estimates are presented in table 8.

We also spoke with NMFS and Marine Mammal Commission officials to identify the potential impacts of using unreliable information to determine human-caused mortality or maximum removal levels.

In some cases, we found potentially conflicting information within individual stock assessment reports about whether fishery-related mortality was unknown or estimated as zero. In these cases, we used the information that NMFS provided in stock assessment report summary tables to resolve the inconsistencies within the individual stock assessment reports because we considered these estimates to be the agency's final decision. In all cases, these summary tables identified the fishery-related mortality estimates for these stocks as zero rather than unknown.

To determine the extent to which NMFS has established take reduction teams for those marine mammal stocks that meet the statutory requirements, we analyzed stock assessment reports for 1995 through 2007 and lists of fisheries for 1996 through 2008 and identified marine mammal stocks that met the statutory requirements for establishing take reduction teams. To do this, we reviewed the MMPA and identified the statutory requirements for establishing take reduction teams, then interviewed officials from the National Oceanic and Atmospheric Administration's (NOAA) Office of General Counsel to verify that we had identified the correct requirements. We also analyzed the stock assessment reports and list of fisheries and identified all of the stocks that have met the statutory requirements, which include marine mammal stocks that (1) were listed as strategic according to a final stock assessment report and (2) interacted with a Category I or II fishery according to a current list of fisheries. We developed a database and used it to analyze this information. Once we identified the marine mammal stocks that met the statutory requirements, we verified with NMFS officials the stocks for which the agency has already established take reduction teams. On the basis of this information, we determined which stocks met the statutory requirements but are not covered by a team. We met with NMFS officials to review and verify our findings, and interviewed NMFS officials to obtain reasons why the agency has not established take reduction teams for these stocks. We also met with representatives of the Marine Mammal Commission to review our findings.

To determine the extent to which NMFS has met the MMPA's deadlines for the five take reduction teams subject to the deadlines and the reasons for any delays, we

- identified five key deadlines listed in the MMPA for NMFS and take reduction teams and interviewed officials from NOAA's Office of General Counsel to confirm the deadlines;
- obtained and reviewed documentation, such as take reduction plans, *Federal Register* notices announcing the establishment of teams, and NMFS's proposed and final take reduction plans and implementing regulations published in the *Federal Register*;

- analyzed the dates published in the *Federal Register* documents to determine whether each of the five take reduction teams had met their statutory deadlines; and,
- met with NMFS officials to confirm the accuracy of our analysis of information published in *Federal Register* notices.

To determine the reasons for any delays in meeting the statutory deadlines, we interviewed take reduction team coordinators from NMFS's Office of Protected Resources, officials from NOAA's Office of General Counsel, marine biologists in NMFS's Fishery Science Centers, and members of each of the five teams subject to the deadlines. We also obtained and reviewed NMFS documentation about various laws and executive orders that the agency must comply with when publishing proposed and final take reduction plans in the *Federal Register*.

To determine the extent to which NMFS has developed a comprehensive strategy for evaluating the effectiveness of the take reduction plans that have been implemented, we reviewed the MMPA and relevant NMFS documentation and spoke with NMFS officials and Scientific Review Group chairs regarding the (1) performance goals used by NMFS to assess the success of take reduction regulations, (2) actions taken prior to implementing proposed regulations to increase the likelihood that the regulations will achieve these performance goals, and (3) extent to which NMFS has information about fisheries' compliance with implemented take reduction regulations. We also reviewed stock assessment reports from 1995 through 2007 for stocks covered by three of the four take reduction teams with final regulations in place to determine whether the stocks met the short- and long-term goals in each of those years.[56] To calculate whether the goals were met prior to implementation of the take reduction regulations, we used the last year for which the fishery-related mortality estimates in the stock assessment reports did not include any information about incidental take that was collected after the regulations were implemented. We excluded the strategic bottlenose dolphins from our review due to methodological differences between the way NMFS reports on fishery-related mortality and maximum removal levels for them versus for the other stocks. Specifically, due to concerns about the stock definition for the Western North Atlantic coastal bottlenose dolphins covered by the Bottlenose Dolphin take reduction team, NMFS further divides this population into management units. NMFS identifies different fishery-related mortality estimates for each of these management units, but not for the Western North Atlantic coastal bottlenose dolphins as a whole, making it difficult to determine whether the total population met the short- and long-term goals. In addition, we assessed the reliability of the data used to determine whether NMFS has met the goals for the strategic stocks covered by three of the four take reduction teams with final regulations. To do this, we analyzed the extent to which key data elements met data quality criteria identified by the MMPA and NMFS. We reviewed strategic stocks because they are most likely to be at continued risk of fishery-related take leading to unsustainable population declines. We also compared the data for the year prior to when the regulations were first implemented with the data from 2007 to identify any changes that occurred in meeting the goals before and after the take reduction regulations went into effect.

We conducted this performance audit from September 2007 to December 2008 in accordance with generally accepted government auditing standards. Those standards require that we plan and perform the audit to obtain sufficient, appropriate evidence to provide a reasonable basis for our findings and conclusions based on our audit objectives. We believe that the evidence obtained provides a reasonable basis for our findings and conclusions based on our audit objectives.

APPENDIX II. COMMERCIAL FISHING TECHNIQUES AND HOW MARINE MAMMALS ARE AFFECTED

The table below presents information about select commercial fishing techniques, including the type of gear, how the injury or death occurs, and examples of marine mammals affected.

Table 9. Commercial Fishing Techniques and How Marine Mammals Are Affected

Type of fishery	Type of gear	How injury or death occurs	Examples of marine mammals affected
Gillnet	A gillnet is a curtain of netting that hangs in the water at various depths, suspended by a system of floats and weights. Gillnets may sometimes be anchored. The mesh spaces are large enough for a fish's head to pass through, but not its body. As fish, such as sardines, salmon, or cod try	Marine mammals get entangled in the nets or fishing lines associated with the gear.	Dolphins (bottlenose and common) Large whales (right, humpback, and sperm) Harbor porpoises

	to back out, their gills are entangled in the net or buoy lines.		
Longline	Longline fishing is conducted by extending a central fishing line behind a fishing boat that ranges from 1 to more than 50 miles long. This central line is strung with smaller lines of baited hooks, which hang at spaced intervals. After leaving the line to soak for a time to attract fish, fishermen return to haul in their catch, such as tuna or swordfish.	Marine mammals are attracted to the baited hooks or the catch and become caught on the hooks or the catch on the hooks. They might also come into accidental contact with gear and become entangled in the fishing gear.	Dolphins (Risso's) Small whales(pilot and false killer)
Long-haul and beach-seine	Long-haul seine fishing uses very large nets (approximately 3,000 to 6,000 feet) pulled by two boats that encircle fish, such as bluefish and croaker, and are then gathered together around a fixed stake. Beach seine fishing involves setting large nets in the water near a beach with the top floating on the surface and bottom falling deeper in the water. The nets are then pulled up onto the beach, entrapping fish in their path.	Marine mammals can get entangled in the large nets that encircle fish.	Dolphins (bottlenose)
Stop/pound nets	Stop net fishing uses a stationary, anchored net extended perpendicular to the beach. Once the catch accumulates near the end of the stop net, a beach haul seine is used to capture fish and bring them ashore. The stop net is traditionally left in the water for 1–5 days, but can be left as long as 15 days. Stop nets are used to catch mullet. Pound nets are stationary gear in nearshore coastal and estuarine waters. Pound net gear includes a large mesh lead posted perpendicular to the shoreline and extending outward to the corral, or "heart," where the catch accumulates. Pound nets typically catch weakfish, spot, and croaker.	Marine mammals can get entangled in the stationary nets along with the fish the nets intend to catch.	Dolphins (bottlenose)

Table 9. (Continued)

Type of fishery	Type of gear	How injury or death occurs	Examples of marine mammals affected
Traps/pots	Traps and pots are submerged cages that usually lie on the ocean floor, attract fish or shellfish, and hold them alive until fishermen return to haul in the gear. Ropes run between the cages along the ocean floor and to a buoy floating at the surface, so fishermen can locate their gear.	Marine mammals get entangled in the rope connecting the cages to each other and the floating buoy. Specifically, right whales feed with their mouths open for extended periods of time and can become entangled in rope and other gear.	Large whales (right, humpback, and fin) Dolphins (bottlenose)
Trawl	Trawlers tow a cone-shaped net behind a fishing boat. They tow nets at various depths, ranging from just below the surface to along the ocean floor, depending on the type of fish they are trying to catch.	Marine mammals become entangled or caught within the nets.	Dolphins (common and white-sided) Small whales (pilot)

Source: GAO analysis of data from the National Marine Fisheries Service and the Monterey Bay Aquarium.

End Notes

[1] These five marine mammals are (1) Western Atlantic stock of North Atlantic right whales, (2) Gulf of Maine stock of humpback whales, (3) Gulf of Maine/Bay of Fundy stock of harbor porpoises, (4) California stock of long-beaked common dolphins, and (5) Hawaii stock of false killer whales.

[2] Under 16 U.S.C. § 1387(f)(3), if there is insufficient funding available to develop and implement a take reduction plan for all stocks that meet the requirements, the Secretary of Commerce must establish teams according to the priorities listed in the statute. Further, under 16 U.S.C. § 1387(f)(3)(6)(A), the Secretary has the discretion to establish take reduction teams for any marine mammal stock that interacts with a Category I fishery and for which the Secretary has determined, after notice and opportunity for public comment, has a high level of mortality and serious injury across a number of such marine mammal stocks.

[3] Optimum sustainable population is defined by the MMPA as "with respect to any population stock, the number of animals which will result in the maximum productivity of the population or the species, keeping in mind the carrying capacity of the habitat and the health of the ecosystem of which they form a constituent element."

[4] Specifically, a fishery is classified as Category I if it is by itself responsible for the annual removal of 50 percent or more of any stock's maximum removal level. A fishery is classified as Category II if it is one that, collectively with other fisheries, is responsible for the annual removal of more than 10 percent of any marine mammal stock's maximum removal level and that is by itself responsible for the annual removal of between 1 and 50

percent, exclusive, of any stock's maximum removal level. Category III fisheries have a remote likelihood of, or no known incidental mortality and serious injury of marine mammals. Specifically, Category III fisheries include, among others, those that collectively with other fisheries are responsible for the annual removal of 10 percent or less of any marine mammal stock's maximum removal level.

[5] If a strategic stock has human-caused mortality and serious injury that is less than the maximum removal level and the stock interacts with Category I or II fisheries, this deadline is 11 months instead of 6 months. The deadline is also 11 months for nonstrategic stocks interacting with Category I fisheries that NMFS has found, after notice and public comment, to have a high level of mortality across a number of marine mammal stocks.

[6] If a take reduction team addressing a strategic stock whose human-caused mortality and serious injury is above the maximum removal level does not submit a draft plan to NMFS within 6 months, NMFS must publish a proposed plan within 8 months of the team's establishment. For strategic stocks whose human-caused mortality and serious injury is below the maximum removal level but that interact with Category I or II fisheries, NMFS's deadline is 13 months.

[7] One team—the Atlantic Trawl Gear—is not subject to the statutory deadlines. NMFS established the Atlantic Trawl Gear take reduction team as a result of a settlement agreement ending the 2002 lawsuit brought by an environmental group. At the time of the settlement agreement, the stocks covered by the team were strategic and interacting with Category I fisheries. After conducting the research and surveys the settlement required, NMFS realized that the stocks were not strategic. NMFS chose to abide by the settlement agreement's requirement to establish the team despite this change in the strategic status because the stocks were interacting with a Category I fishery. The MMPA gives NMFS the discretion to establish teams for nonstrategic stocks interacting with Category I fisheries that NMFS has determined, after notice and public comment, to have a high level of mortality across a number of marine mammal stocks. However, at the present time, the fisheries involved are no longer Category I fisheries and NMFS has never made the required determination. The MMPA is silent on deadlines for teams, such as the Atlantic Trawl Gear team, that address nonstrategic stocks that do not interact with Category I fisheries. Therefore none of the deadlines apply to this team.

[8] The Marine Mammal Commission is composed of three presidential appointees who are knowledgeable in the fields of marine ecology and resource management and are not in a position to profit from the taking of marine mammals.

[9] The MMPA divides jurisdiction over marine mammals between the U.S. Fish and Wildlife Service and NMFS but gives NMFS the exclusive authority to establish take reduction teams and implement take reduction plans for all marine mammals. NMFS has not established take reduction teams for any of the marine mammals under the Fish and Wildlife Service's jurisdiction (sea otters, polar bears, manatees, dugongs, and walruses). This chapter focuses on marine mammals under NMFS's jurisdiction.

[10] Under 16 U.S.C. § 1387(f)(3), highest priority must be given to the development and implementation of take reduction plans for species or stocks whose level of incidental mortality and serious injury exceeds the maximum removal level, those that have a small population size, and those that are declining most rapidly.

[11] A stock is also considered strategic if it is designated as depleted under the MMPA or if it is listed or likely to be listed as endangered or threatened under the Endangered Species Act. For these stocks, human-caused mortality does not necessarily have to exceed the maximum removal level.

[12] The minimum population estimate is an estimate of the number of animals in a stock that (1) is based on the best available scientific information on abundance, incorporating the precision and variability associated with such information, and (2) provides reasonable assurance that the stock size is equal to or greater than the estimate.

[13] Regional Scientific Review Groups were established by the 1994 amendments to the MMPA. The MMPA directs NMFS to identify members of these groups in consultation with the Marine Mammal Commission, among others.

[14] For the purposes of categorizing fisheries, NMFS uses only estimates for fishery-related mortality rather than the estimates for total human-caused mortality.

[15] Fisheries are classified as Category I if the fishery by itself is responsible for the annual removal of 50 percent or more of any stock's maximum removal level. A Category II fishery is one that, collectively with other fisheries, is responsible for the annual removal of more than 10 percent of any marine mammal stock's maximum removal level, and by itself is responsible for the annual removal of between 1 and 50 percent, exclusive, of any stock's maximum removal level.

[16] Specifically, the MMPA requires the Secretary to publish the take reduction plan proposed by the team, any changes proposed by the Secretary with an explanation of the reasons for the changes, and proposed regulations to implement such a plan.

[17] "Small entities" includes businesses, small governmental jurisdictions, and certain not-for-profit organizations.

[18] Exec. Order No. 12866, 58 Fed. Reg. 51735 (Sept. 30, 1993); Exec. Order No. 13,132, 64 Fed. Reg. 43255 (Aug. 4, 1999); Exec. Order No. 12898, 59 Fed. Reg. 7629 (Feb. 11, 1994); Exec. Order No. 13158, 65 Fed. Reg. 34909 (May 26, 2000).

[19] The MMPA directs NMFS to establish take reduction teams for stocks designated as strategic that interact with Category I or II fisheries.

[20] NMFS has identified a total of 156 marine mammal stocks in United States waters that are under its jurisdiction. However, a NMFS scientist told us that additional marine mammals exist in the waters off the Pacific islands under NMFS's jurisdiction that have not been identified and defined as stocks because the agency does not have the necessary data for these marine mammals. Our analysis of the stock assessment reports focused on 113 of the 156 stocks that NMFS has identified, excluding from our analysis the 19 stocks that are currently addressed by take reduction teams and the 24 that are designated as strategic because of ESA listing or MMPA designation and therefore do not rely on human-caused mortality estimates and maximum removal levels to determine strategic status.

[21] NOAA Fisheries National Task Force for Improving Marine Mammal and Turtle Stock Assessments, *A Requirements Plan for Improving the Understanding of the Status of U.S. Protected Marine Species.* National Marine Fisheries Service, NOAA Technical Memorandum NMFS-F/SPO-63 (Silver Spring, Maryland: September 2004).

[22] We analyzed a stratified random sample of 28 stocks out of the 74 stocks that had both mortality estimates and maximum removal levels and used the results from this sample to estimate the results for all 74 stocks. For this reason, numbers in the report about these 74 stocks are described as approximations. We calculated 95 percent confidence intervals for each of the estimates made from our sample. The confidence intervals for these estimates are presented in appendix I, table 8.

[23] NMFS guidelines identified data that are 8 years old or older as unreliable because a population that declines at 10 percent per year from a sustainable level would be reduced to less than 50 percent of its original abundance in 8 years.

[24] National Marine Fisheries Service, Office of Protected Resources, *Review of Commercial Fisheries' Progress Toward Reducing Mortality and Serious Injury of Marine Mammals Incidental to Commercial Fishing Operations.* United States Department of Commerce, National Oceanic and Atmospheric Administration (Silver Spring, Maryland: 2004).

[25] NMFS calculates precision by identifying a coefficient of variation (CV) for each estimate. The lower the CV percentage, the more precise the estimate. NMFS's publications state that CVs of 30 percent or lower are considered to have a desirable level of precision appropriate for determining strategic status. Therefore, estimates with CVs greater than 30 percent are less precise than NMFS guidelines recommend.

[26] Scientists may use data sources—such as stranding data—that do not allow them to make statistically based estimates of total fishery-related mortality. A stranded marine mammal is either dead and on the beach or shore, or in the water, or is alive and on the beach or shore but unable to return to the water under its own power. Information from such sources is anecdotal because it is not based on scientific sampling techniques that are used to make generalizable estimates.

[27] According to NMFS officials, the National Observer Program uses the following categories to characterize the observer coverage level for fisheries: none, pilot, baseline, near-adequate, or adequate coverage. NMFS officials stated that the agency often chooses not to observe trap/pot fisheries because the nature of marine mammal interactions with this type of fishing gear make it unlikely that an observer would see any instances of incidental take.

[28] NOAA Fisheries National Task Force for Improving Marine Mammal and Turtle Stock Assessments, A Requirements Plan for Improving the Understanding of the Status of U.S. Protected Marine Species.

[29] The two take reduction teams no longer in existence were the Atlantic Offshore Cetacean and Mid-Atlantic take reduction teams. Some of the stocks covered by the Atlantic Offshore Cetacean team are now covered by the Pelagic Longline and Atlantic Trawl Gear teams. The Mid-Atlantic take reduction team merged with the Gulf of Maine Harbor Porpoise team, and both teams covered the same stock of harbor porpoises. The merged team is now referred to by NMFS as the Harbor Porpoise team.

[30] According to NMFS's 2007 stock assessment reports.

[31] In the Gulf of Mexico, the two strategic stocks are the Northern Gulf of Mexico coastal stock and the Gulf of Mexico bay, sound, and estuarine stock of bottlenose dolphins. In the waters off Alaska's coast, the six strategic stocks are the Gulf of Alaska, Bering Sea, and Southeast Alaska stocks of harbor porpoises; the Western and Eastern U.S. stocks of Steller sea lions; and the Western North Pacific stock of humpback whales.

[32] The Gulf of Mexico Gillnet fishery and the Gulf of Mexico Menhaden Purse Seine fishery.

[33] Sperm whales meet the statutory requirements for establishing a team because they are strategic and interact with a Category I fishery, but this fishery is a Category I fishery because of its incidental take of other marine mammal stocks.

[34] As explained in footnote 7, the Atlantic Trawl Gear team is not subject to the MMPA's deadlines.

[35] From 1996 to 2001, these whales were covered by a former take reduction team known as the Atlantic Offshore Cetacean team, but this team was disbanded in 2001 without the publication of a final take reduction plan.

[36] According to the *Federal Register* notice announcing the establishment of the Mid-Atlantic take reduction team, the team was not established to address bottlenose dolphins. At preestablishment meetings, NMFS and the team determined that there was not enough information available about the bottlenose dolphins to implement a take reduction plan at that time and agreed to delay establishing a take reduction team and developing a take reduction plan specific to bottlenose dolphins until more information was obtained. However, the team provided NMFS with research and data recommendations that addressed bottlenose dolphins in its 1997 draft take reduction plan.

[37] The three teams are the Atlantic Large Whale, Harbor Porpoise, and Pacific Offshore Cetacean.

[38] Because the deadline for publication of the Pelagic Longline final plan occurred after we had concluded our audit work, we did not interview NMFS to ascertain the reasons for the delay.

[39] For example, see GAO, *The Results Act: An Evaluator's Guide to Assessing Agency Annual Performance Plans,* GAO/GGD-10.1.20 (Washington, D.C.: April 1998).

[40] The long-term goal is also known as the zero mortality rate goal (ZMRG) or reducing incidental take to an insignificant level approaching a zero mortality and serious injury rate. The goal of commercial fisheries reducing mortality and serious injuries to insignificant

levels approaching a zero mortality rate goal within 5 years of a take reduction plan's implementation must take "into account the economics of the fishery, the availability of existing technology, and existing state or regional fishery management plans." The MMPA also has a deadline of April 30, 2001, for "commercial fisheries to reduce mortality and serious injuries to insignificant levels approaching a zero mortality rate goal." The MMPA does not define ZMRG, but NMFS has defined "insignificance threshold" as the default target level of mortality and serious injury for all marine mammal stocks. Under NMFS's regulation, take reduction plans and implementing regulations are the mechanisms that help Category I and II fisheries meet the insignificance threshold but these take reduction plans and regulations must take into account the fishery's economics, availability of existing technology, and existing fishery management plans.

[41] These are the California/Oregon/Washington stock of long-beaked common dolphins and what is now called the Gulf of Maine stock of humpback whales.

[42] In some cases, strategic stocks could be meeting the goal of reducing fishery-related mortality to below the maximum removal level prior to implementation. Specifically, this might be the case for stocks that receive their strategic status determination through an ESA listing or designation as depleted under the MMPA.

[43] While NMFS's guidance provides that the long-term goal must take into account the economics of the fishery, the availability of existing technology, and existing state or regional fishery management plans, the agency has not specified how it considers these factors in establishing long-term goals for the current take reduction plans. Therefore we examined whether or not the long-term goals had been met by assessing whether fishery-related mortality was less than 10 percent of the maximum removal level. We used data from 2007 to measure whether the goals had been met rather than measuring 5 years after the implementation of each plan's regulations in order to make general determinations about whether these goals are adequate measures of success.

[44] These are the Western North Atlantic stock of fin whales and the Canadian East Coast stock of minke whales. NMFS told us that because minke whales are not a strategic stock, they are not relevant in assessing NMFS's achievement of the long-term goal. However, the Atlantic Large Whale take reduction plan states that a goal of the plan is to reduce entanglement-related serious injury of minke whales to insignificant levels approaching zero mortality and serious injury rate. Thus, NMFS has stated its intent in the plan to achieve the long-term goal for the minke whales.

[45] The two stocks that had already met the short-term goal were the Western North Atlantic stock of fin whales and the California/Oregon/Washington stock of fin whales. The two stocks that had already met the long-term goal were the California/Oregon/Washington stock of fin whales and the California/Oregon/Washington stock of short-beaked common dolphins.

[46] There are currently 13 stocks—10 strategic and 3 nonstrategic—covered by take reduction regulations. According to the MMPA, as amended, the short-term goal is applicable only to strategic stocks. Under authority granted by the MMPA, NMFS may choose to establish teams for nonstrategic stocks, but these stocks are subject only to the long-term goal of reducing fishery-related mortality to 10 percent of the maximum removal level. However, by definition, these stocks have already met the short-term goal.

[47] Western North Atlantic coastal bottlenose dolphins are also strategic. We chose not to assess progress by the Bottlenose Dolphin take reduction team in meeting the goals due to the unique data collection system that NMFS uses for this team. Specifically, due to concerns about the stock definition for the Western North Atlantic coastal bottlenose dolphins covered by the Bottlenose Dolphin take reduction team, NMFS further divides this population into management units. NMFS identifies different fishery-related mortality estimates for each of these management units, but not for the Western North Atlantic coastal bottlenose dolphins as a whole, making it difficult to determine whether the total population met the short- and long-term goals.

[48] The humpback whale is listed as an endangered species under the ESA and therefore is designated as strategic even though human-caused mortality was lower than the maximum removal level when the take reduction team was established.

[49] Marjorie C. Rossman and Debra L. Palka, "A Review of Coastal Bottlenose Dolphin Bycatch Mortality Estimates in Relation to the Potential Effectiveness of the Proposed CBDTRP." Northeast Fisheries Science Center Protected Species Branch (Woods Hole, Massachusetts: 2004).

[50] Office of Protected Resources, National Marine Fisheries Service, *Harbor Porpoise Take Reduction Plan (HPTRP) Final Environmental Assessment and Final Regulatory Flexibility Analysis* (Silver Spring, Maryland: 1998).

[51] Fishery observer programs place individuals on commercial fishing vessels to observe operations, including documenting any instances of incidental take of marine mammals.

[52] Debra Palka, "Effect of Pingers on Harbor Porpoise and Seal Bycatch." Northeast Fisheries Science Center (Woods Hole, Massachusetts: 2007).

[53] GAO, *National Marine Fisheries Service: Improved Economic Analysis and Evaluation Strategies Needed for Proposed Changes to Atlantic Large Whale Protection Plan,* GAO-07-881 (Washington, D.C.: July 20, 2007).

[54] Maximum removal level is defined as the maximum number of animals—not including natural mortalities—that may be removed from a marine mammal stock while allowing that stock to reach or maintain its optimum sustainable population.

[55] National Marine Fisheries Service, *Guidelines for Preparing Stock Assessment Reports Pursuant to Section 117 of the Marine Mammal Protection Act: SAR Guidelines, June 2005 Revision.* (Silver Spring, Maryland: June 2005), and NOAA Fisheries National Task Force for Improving Marine Mammal and Turtle Stock Assessments, *A Requirements Plan for Improving the Understanding of the Status of U.S. Protected Marine Species.* National Marine Fisheries Service, NOAA Technical Memorandum NMFS-F/SPO-63 (Silver Spring, Maryland: September 2004).

[56] Three stocks, the Canadian East Coast stock of minke whales, the California/Oregon/Washington stock of northern right whale dolphins, and the California/Oregon/Washington stock of short-beaked common dolphins, were included in the teams even though they were not strategic when the teams were established. In accordance with the MMPA, only the long-term goal applies to these stocks, so we did not analyze whether these stocks had met the short-term goal.

CHAPTER SOURCES

The following chapters have been previously published:

Chapter 1 – This is an edited, excerpted and augmented edition of a United States Congressional Research Service publication, Report Order Code RL30120, dated June 11, 2007.

Chapter 2 – This is an edited, excerpted and augmented edition of a United States Government Accountability Office (GAO), Report to the Ranking Member, Subcommittee on Oceans, Atmosphere, Fisheries and Coast Guard, Committee on Commerce, Science, and Transportation, U.S. Senate. Publication GAO-07-881, dated July 2007.

Chapter 3 – This is an edited, excerpted and augmented edition of a United States Government Accountability Office (GAO), Report to the Chairman, Committee on Natural Resources, U.S. House of Representatives. Publication GAO-09-78, dated December 2008.

INDEX

A

abusive, 25
accidental, 16, 71, 163
accountability, 42
accuracy, 29, 56, 122, 132, 133, 161
achievement, 152, 168
acoustic, 21, 34, 35, 36, 52, 57, 122, 141, 149, 153
acoustical, 15
acute, 13
administration, 11, 37, 44, 45, 64, 74
administrative, 31, 37, 53, 56
adult, 24, 53
advocacy, 14
afternoon, 59
age, 15, 56, 158
agent, 59
aggregation, 74
aid, 51
air, 25
air pollution, 25
algae, 92
alternative, 22, 23, 39, 52, 66, 68, 72, 73, 83, 89
alternatives, 66, 73, 74, 78, 112
amendments, ix, 2, 3, 8, 9, 10, 15, 16, 19, 20, 23, 26, 33, 38, 42, 44, 45, 46, 50, 57, 59, 71, 115, 118, 119, 125, 128, 129, 166
analysts, 50

Animal and Plant Health Inspection Service, vii, viii, 2, 11
animal care, 58
animal health, 19
animal husbandry, 19
animal welfare, 58
animals, ix, 5, 6, 11, 12, 17, 20, 21, 22, 23, 24, 25, 26, 27, 28, 29, 30, 32, 34, 35, 39, 40, 42, 45, 48, 51, 52, 53, 54, 55, 56, 57, 111, 115, 119, 152, 164, 165, 169
annual review, 15
anthropogenic, 34, 35, 36, 58
antibiotic, 22
APHIS, 11, 19, 22, 23, 24, 25, 26, 27, 46, 53, 54, 56
appendix, 67, 70, 96, 98, 101, 106, 144, 166
application, 20, 37, 58
appointees, 165
appropriations, vii, viii, 1, 2, 6, 29, 45, 46, 47
aquaculture, 41
assumptions, 13, 62, 69, 87, 97, 100, 109
auditing, 67, 100, 121, 162
authority, viii, 2, 6, 7, 12, 15, 16, 17, 18, 19, 24, 30, 31, 32, 38, 44, 51, 53, 54, 55, 57, 58, 59, 145, 165, 168
availability, 6, 39, 168
avoidance, 41

B

back, 94, 162
barriers, 39
barter, 51
batteries, 57
behavior, 18, 21, 52
behavioral change, 58
beliefs, 32
benefits, 16, 21, 31, 35, 41, 57
bias, 58, 158
biodiversity, 4
birds, 22, 34
birth, 54
birth rate, 54
blame, 12, 17
blood, 22
boats, 52, 59, 64, 82, 94, 119, 163
bonding, 27
breathing, 64
breeding, 20, 24, 25, 55, 71
bureaucracy, 55

C

calving, 71, 80
candidates, 24
certificate, 22
citizens, 7, 33
classes, 35
Clean Water Act, 54
closure, 129
clubbing, 6
Coast Guard, 46, 63, 154, 171
coatings, 112
coding, 91
coefficient of variation, 117, 166
collision avoidance, 41
collisions, 75, 126
colors, 91
commerce, 39
communication, 28

communities, viii, 14, 29, 30, 31, 32, 38, 39,
    44, 46, 56, 61, 63, 65, 66, 69, 84, 88, 89,
    100, 106, 109, 112
community, 4, 10, 13, 19, 20, 21, 22, 25, 30,
    31, 33, 35, 38, 89, 95, 96, 107, 112
competition, 12, 58
complexity, 8, 9
compliance, viii, 2, 11, 35, 42, 44, 53, 62,
    63, 66, 67, 69, 70, 84, 85, 88, 93, 95, 97,
    98, 100, 107, 109, 110, 117, 122, 124,
    150, 154, 155, 156, 161
components, 64
composition, 56
concrete, 13, 26
confidence, 24, 86, 132, 136, 159, 166
confidence interval, 159, 166
confidence intervals, 159, 166
conflict, 3, 18, 32, 40, 57, 58
conflict of interest, 58
confusion, 27, 32
consensus, 62, 123, 127, 142, 144
conservation, 4, 6, 7, 10, 36, 44, 49, 58, 67,
    71, 72, 99, 107, 113, 125, 140
consolidation, 132
constituent groups, 6
constraints, 45, 122, 130, 135, 140
construction, 56
consulting, 57, 73, 79
consumption, 30, 51
contaminants, 14
contamination, 58
contracts, 29
control, 40
Convention on International Trade in
    Endangered Species, 39
copepods, 71
cost-effective, 57
costs, viii, 14, 20, 23, 24, 31, 61, 62, 66, 68,
    70, 74, 75, 84, 85, 86, 88, 94, 95, 96, 97,
    98, 100, 107, 109, 112
courts, 53
covering, 27, 50
crab, 111
critical habitat, 73, 91
criticism, 45

CRS, 47, 50, 51, 53, 57, 59
crustaceans, 111
cultural heritage, 31
cultural identity, 56
cultural values, 32
culture, 5
cycles, 52

D

danger, 54
data analysis, 30
data collection, 6, 136, 168
database, 78, 111, 112, 160
death, vii, viii, 23, 35, 41, 55, 61, 64, 74, 80,
    117, 123, 125, 141, 144, 162, 164
deaths, 23, 55, 65, 77
decisions, 17, 49, 122, 130, 131, 135, 158
definition, 10, 11, 12, 13, 14, 26, 35, 38, 42,
    43, 48, 59, 121, 131, 152, 161, 168
degradation, 5, 40
Department of Agriculture, 11, 18
Department of Commerce, 11, 48, 95, 101,
    110, 111, 166
Department of Defense, 51
Department of Interior, 111
Department of State, 50
Department of the Interior, 48, 73
deterrence, 15, 16
deterrence regulations, 15
diet, 56
disabled, 54
disaster, 20
discharges, 22, 54
discriminatory, 37
disposition, 55
distribution, 44, 80, 157
diversity, 3, 24
diving, 34
division, 6, 43
download, 57
draft, 10, 50, 62, 65, 70, 74, 95, 96, 98, 110,
    117, 119, 123, 125, 127, 129, 139, 142,
    143, 144, 145, 146, 147, 156, 165, 167
durability, 82

duration, 112, 133
duties, 45

E

ecological, 5, 10, 29, 32, 52
ecologists, 12, 13
ecology, 32, 165
economic disadvantage, 57
economic losses, 16
economic stability, 5
economics, 54, 168
ecosystem, 10, 11, 12, 14, 29, 41, 43, 44,
    51, 52, 164
ecosystems, 3, 6, 11
effluent, 23, 54
effluents, 22
elephants, 41
email, 156
employment, 88, 89, 112
Endangered Species Act, 8, 33, 57, 63, 64,
    117, 118, 128, 158, 165
endocrine, 44
energy, 47
entanglement, 14, 64, 66, 67, 68, 71, 72, 75,
    76, 77, 78, 79, 80, 84, 90, 91, 92, 97, 99,
    111, 168
entanglements, vii, viii, 61, 62, 63, 65, 67,
    69, 71, 73, 74, 76, 77, 79, 90, 91, 92, 93,
    94, 99, 111
entertainment, 50
entrapment, 18
environment, 25, 26, 49, 51, 54, 56, 93
environmental conditions, 25
environmental effects, 25, 128
environmental impact, 36, 57, 65, 66, 74,
    95, 98, 128
environmental protection, 4
Environmental Protection Agency, 54
environmentalists, 24, 31, 34, 36, 37, 38, 40
EPA, 54
estimating, 132
estuarine, 138, 163, 167
ethics, 32
evolution, viii, 2

examinations, 22, 55
exercise, 15, 31, 58
expertise, 19, 53, 72, 127
exploitation, 39, 40
explosives, 34
exporter, 22
exports, 21
extinction, 6, 8
eye, 31

F

failure, 44, 59, 93, 143, 155
family, 51
farms, 15, 52
fear, 20, 23, 29, 39, 40, 56
federal courts, 53
federal government, 7, 10, 12, 15, 24, 28,
     29, 42, 52, 53, 57, 119
federal law, 49
Federal Register, 37, 58, 65, 119, 120, 127,
     129, 137, 138, 141, 145, 146, 147, 148,
     149, 150, 160, 161, 167
feedback, 81, 82
feeding, 27, 52, 56, 71, 80
fees, 10
feet, 68, 75, 83, 89, 91, 92, 113, 163
females, 32
finance, 44, 75
financial resources, 28
financing, 26
fines, 42
fish, 4, 7, 12, 13, 15, 16, 17, 34, 40, 49, 51,
     52, 53, 58, 64, 71, 82, 89, 113, 118, 147,
     162, 163, 164
Fish and Wildlife Service (FWS), vii, viii, 2,
     6, 10, 49, 50, 52, 59, 73, 128, 165
float, 84, 111
floating, 62, 64, 66, 68, 72, 74, 75, 81, 83,
     84, 85, 86, 94, 163, 164
fluctuations, 51
focusing, 5
food, 43, 51, 52, 71
foreign nation, 20, 32
freedom, 26

Freedom of Information Act, 23
freshwater, 111
fuel, 51
funding, viii, 2, 6, 10, 14, 16, 18, 28, 33, 41,
     42, 44, 45, 48, 52, 56, 58, 59, 75, 94,
     113, 116, 122, 125, 127, 130, 131, 133,
     134, 135, 138, 139, 140, 141, 164
funds, 9, 12, 14, 18, 34, 41, 42, 44, 45, 56,
     75, 116, 122, 136, 139
FWS, 6, 8, 9, 11, 13, 18, 19, 20, 22, 23, 28,
     29, 30, 31, 32, 33, 35, 36, 37, 43, 44, 45,
     46, 53, 54, 59

G

gas, 7, 34
gas exploration, 7, 34
gastrointestinal, 43
gastrointestinal tract, 43
genetic diversity, 24
goals, 3, 5, 117, 123, 150, 151, 152, 153,
     155, 156, 161, 168
governance, 47
government, iv, 7, 10, 12, 15, 19, 20, 24, 25,
     28, 29, 31, 37, 39, 40, 42, 48, 52, 53, 57,
     67, 100, 107, 119, 121, 162
Government Accountability Office (GAO),
     171
Government Performance and Results Act,
     123, 150
grants, 10, 12, 29
Greenland, 58
gross national product, 4
groups, 3, 4, 5, 7, 11, 14, 16, 17, 19, 23, 28,
     32, 39, 42, 43, 46, 50, 51, 53, 54, 55, 58,
     67, 72, 76, 79, 107, 108, 109, 111, 119,
     166
growth, 126
guidance, 35, 42, 54, 112, 122, 130, 133,
     134, 159, 168
guidelines, 14, 25, 26, 59, 126, 132, 133,
     134, 152, 158, 159, 166

H

habitat, 4, 5, 11, 27, 29, 40, 52, 53, 73, 91, 133, 143, 164
handling, 19
harassment, viii, 2, 5, 8, 11, 15, 27, 34, 35, 36, 37, 38, 42, 48, 59
harm, 4, 15, 34, 49, 52, 72, 112, 155
harvest, 12, 13, 14, 29, 30, 31, 32, 39, 51, 56, 57, 58, 112
harvesting, 13
hazards, 62, 81, 83
health, 5, 11, 19, 22, 25, 28, 41, 47, 56, 58, 164
healthcare, 24
hearing, 3, 21, 46, 52
heart, 163
height, 84
herring, 16
high pressure, 36
high-speed, 34, 35
host, 12, 13
house, vii, 1, 3, 46, 47, 52, 53, 59, 60, 117, 171
household, 56
human actions, 6
human behavior, 52
humane, 19, 45, 53
humans, 22, 27, 46, 52, 55, 59
hunting, 30, 31, 33, 56, 57, 126
husbandry, 19, 20, 24

I

identification, 111
identity, 56
immunology, 21
implementation, 8, 19, 45, 46, 53, 62, 66, 96, 98, 107, 121, 124, 127, 150, 151, 152, 161, 165, 168
imports, 8, 21, 33, 57, 58
in situ, 30
inbreeding, 24
incentive, 13, 16, 55

indication, 154
indices, 52
indigenous, 5
individual action, 35
industrial, 14
ineffectiveness, 63, 70, 90, 93, 153
infections, 22
inflation, 112
Information Quality Act, 128
information sharing, 53
innovation, 12
inspection, 53
inspection, vii, viii, 2, 11, 20, 24
inspectors, 23, 53
institutions, 23, 55, 77
insurance, 20, 27
integration, 17, 55
integrity, 5
interaction, 16, 17, 27, 32, 42, 56, 141
interactions, viii, 1, 8, 9, 11, 14, 15, 17, 18, 21, 38, 40, 41, 46, 51, 52, 119, 167
interest groups, 16
International Trade, 39
interstate, 51, 72, 119
interval, 15
interviews, 85, 96, 106, 108
invasive, 21, 37, 38, 41
investment, 22
isolation, 55
ivory, 32

J

jobs, 51, 63, 69, 85, 89
judge, 136
judgment, 78, 130
jurisdiction, 3, 7, 8, 10, 19, 48, 53, 112, 119, 126, 165, 166
jurisdictions, 166
justification, 38, 57

K

killing, 5, 17, 41

L

labor-intensive, 146
land, 111
language, 18, 30, 34, 42, 50, 54, 147
law, viii, 2, 19, 21, 32, 41, 49, 100, 155
law enforcement, 100
laws, 50, 128, 146, 148, 155, 161
legislation, 3, 24, 31, 50
LFA, 35
liberal, 17, 57
life expectancy, 25
lifespan, 85, 88, 108, 112
lifestyle, 56
likelihood, 49, 91, 122, 133, 134, 153, 161,
    165
limitations, 94, 109, 116, 117, 121, 124,
    129, 131, 134, 135, 150, 152, 156, 158
links, 72, 74, 79, 80
lithosphere, 35
litigation, 9
lobsters, 88
local government, 40
location, 4, 54, 72, 83, 85, 86, 91, 94
logistics, 23
long-term impact, 53
losses, 16
low risk, 92

M

magnetic, iv
maintenance, 19, 25, 55
mandates, 37, 59, 135
manufacturer, 92
marine environment, 22, 26, 36, 38, 45, 51,
    152
market, 31, 39, 56, 112
marketing, 51
marketplace, 56
markets, 30, 39, 58
measures, vii, ix, 1, 15, 16, 27, 34, 45, 57,
    64, 67, 70, 72, 76, 90, 115, 118, 119,
    127, 144, 147, 153, 156, 168

meat, 30, 32, 71
median, 86
membership, 50
memorandum of understanding, 53
memorandum of understanding (MOU), 53
military, 11, 25, 34, 35, 36, 42
minority, 31, 127
misleading, 109
missions, 47, 48
mixing, 24, 55
models, 13, 31
money, 51
morality, 151
moratorium, 6, 7, 32, 51
morning, 59
mortality rate, 14, 54, 71, 111, 167
mouth, 71, 79
movement, 24, 26, 43, 64
multiplier, 13

N

nation, 3, 20, 54, 57
National Academy of Sciences, 44, 110
National Marine Fisheries Service, vii, viii,
    ix, 2, 6, 49, 51, 53, 58, 59, 61, 63, 64, 98,
    115, 117, 118, 157, 164, 166, 169
National Oceanic and Atmospheric
    Administration, 6, 48, 50, 63, 70, 95,
    107, 110, 117, 118, 160, 166
National Science Foundation, 48
Native American, viii, 1, 3, 5, 11, 28, 29,
    30, 43
natural, 5, 12, 14, 20, 26, 51, 52, 111, 119,
    126, 133, 169
natural resources, 5
negative experiences, 82
negligence, 24, 55
negotiation, 47
noise, viii, 1, 14, 16, 25, 34, 35, 36, 45, 48,
    52, 57, 58
nongovernmental, 99
non-human, 141
normal, 26, 64
not-for-profit, 166

nutrition, 21

O

oat, 164
obligation, 155
observations, 130
oceans, 24, 48, 49, 60
Office of Management and Budget, 44, 117, 128, 148, 149
Offices of Congressional Relations and Public Affairs, 98
offshore, 4, 7, 79, 112
offshore oil, 7
oil, 7, 71
OMB, 117, 128
otters, 6, 17, 18, 29, 53, 165
oversight, viii, 2, 3, 19, 26, 46, 48, 49, 53, 67, 124, 156

P

partnership, 112
pathogenic, 22
pathogens, 22, 55
PCBs, 49
peer, 22, 37, 56, 65
peer review, 22, 65
penalties, 49
periodic, 26
permit, viii, 2, 7, 10, 20, 21, 26, 27, 33, 35, 37, 38, 40, 44, 45, 50, 53, 58, 59, 73, 90, 111, 112
photographs, 90, 112
physical environment, 55
physiological, 34, 52
physiology, 21, 52
plankton, 71
planning, 17, 70, 97, 109
plants, 4, 51
play, 90
poisoning, 6
polar bears, 6, 10, 28, 29, 33, 57, 165
policymakers, 146

pollution, 25
polychlorinated biphenyls (PCBs), 49
pools, 27
poor, 136
population size, 77, 116, 117, 122, 123, 126, 127, 132, 133, 134, 136, 139, 140, 141, 142, 143, 153, 158, 165
ports, 75
power, 15, 167
pragmatic, 6
preferential treatment, 37
pressure, 36, 39, 41, 72
prices, 86, 112
private, 5, 42, 54
private sector, 5
proactive, 26
productivity, 11, 12, 13, 51, 56, 164
profit, 165, 166
program, viii, 2, 10, 11, 18, 19, 26, 28, 29, 30, 33, 41, 43, 45, 47, 48, 53, 59, 74, 75, 95, 124, 131, 134, 135, 150, 154
protected area, 16, 53
protected areas, 16, 53
protein, 41
protocols, 26, 55
proxy, 113
public domain, vii
public education, 50
public funding, 10
public notice, 148
pulses, 57

Q

quotas, 12, 32, 33, 57

R

rain, 35
random, 159, 166
range, vii, viii, 2, 13, 18, 48, 52, 54, 56, 62, 66, 69, 70, 81, 84, 85, 86, 87, 88, 94, 95, 96, 97, 107, 108, 109, 127
reactivity, 25

reality, 84
reasoning, 14
recognition, 36
record keeping, 54
recovery, 7, 13, 14, 18, 32, 39, 44, 52
recreational, 34, 44, 47, 51, 92
recycling, 75
regional, 17, 45, 49, 88, 89, 111, 112, 126, 168
regular, 36, 37
regulation, 19, 27, 35, 36, 44, 55, 56, 66, 73, 74, 75, 87, 88, 89, 91, 93, 97, 98, 109, 112, 150, 168
regulators, 67, 76, 79
regulatory requirements, 63, 69, 93
rehabilitation, 17, 25, 55
relationships, 12, 29, 41, 52, 56
relevance, 55
reliability, 120, 132, 158, 162
replacement rate, 109
reproduction, 41, 48
resolution, 32
resource management, 165
resources, 3, 4, 5, 6, 28, 40, 43, 44, 45, 47, 49, 53, 58, 111, 112, 122, 124, 135, 155
responsibilities, 6, 45, 73, 111
retail, 51
revenue, 69, 88
risk, 14, 27, 48, 64, 66, 67, 68, 69, 71, 72, 74, 75, 76, 78, 79, 84, 92, 99, 106, 112, 162
rocky, 62, 66, 68, 81, 82, 83, 87, 96, 98, 99, 107, 109
rural, 51, 56

S

safeguard, 21, 22, 37, 54, 57
safety, 40, 62, 64, 68, 81, 83
sales, 31
salmon, 47, 58, 118, 162
sample, 120, 122, 132, 133, 158, 159, 166
sampling, 30, 50, 94, 167
SAR, 169
schooling, 16

scientific community, 35
sea urchin, 17
seabirds, 49
seafood, 89
seals, 6, 16, 29, 30, 40, 51, 140
seawater, 71
Secretary of Commerce, 6, 11, 43, 53, 95, 98, 106, 156, 157, 164
seismic, 34, 57
selecting, 66
self-regulation, 56
self-report, 29
sensitivity, 21
series, 53
sex, 15, 56
sharing, 31, 41, 51, 53
shellfish, 51, 58, 164
shelter, 51
shipping, 75
shores, 110
short period, 52, 71, 82
short-term, 4, 55, 58, 124, 151, 152, 168, 169
shrimp, 111
side effects, 16
signals, 52
skills, 53
skin, 32, 56
snaps, 62, 83
social adjustment, 26
social environment, 26
social integration, 55
social isolation, 55
social structure, 55
social systems, 27
socioeconomic, 44
sounds, 57, 149
spatial, 118
specificity, 37
spectrum, 55
speculation, 25, 35
speed, 34, 35, 75
sperm, 52, 137, 140, 162
stability, 5
stakeholder, 65, 112

stakeholders, 65, 66, 91, 92, 95, 96, 106, 108, 109
standards, 19, 20, 23, 25, 27, 37, 54, 56, 57, 58, 67, 100, 121, 162
state regulators, 76, 79
statutes, 8, 9, 39
statutory, vii, viii, 2, 9, 116, 120, 121, 122, 123, 124, 128, 129, 135, 140, 141, 142, 143, 144, 145, 146, 148, 149, 156, 157, 160, 161, 165, 167
strategies, viii, 53, 61, 63, 65, 66, 69, 95, 98, 99, 100, 107
strength, 74, 79, 80
stress, 25, 50
strikes, 34, 41, 75
subjective, 130
subsistence, viii, 1, 5, 7, 13, 28, 29, 30, 31, 32, 41, 46, 51, 56, 57, 58, 126
substances, 49
suffering, 5, 17
summer, 56, 71
supplemental, 36
suppliers, 85, 86, 100
supply, 31
surplus, 25
survival, 7, 18, 39, 48, 64, 94, 110
sustainability, 5, 14
swimmers, 42, 59

T

tanks, 26
team members, ix, 116, 119, 123, 127, 129, 139, 143, 144, 146, 147, 151, 154
temperature, 86
territorial, 87
territory, 31
testimony, 3, 52
threat, 22, 28, 35, 42
threatened, 8, 15, 26, 31, 33, 40, 49, 54, 73, 118, 120, 126, 128, 158, 165
threats, 125
threshold, 15, 89, 168
tides, 79
time consuming, 94

time frame, 151
timing, 15
tissue, 23
title, 108
total costs, 69
toxic, 49
toxic substances, 49
toys, 56
trade, viii, 1, 31, 39, 57, 58
tradition, 5, 32
traffic, 34, 92
training, 24
transfer, 12, 20, 22, 27, 54, 59
translocation, 18
transparency, 29
transport, 19, 34, 53
transportation, 51
traps, 64, 72, 82, 83, 118
travel, 36, 64, 76
trawling, 125
tribal, 51
tribes, 5, 28
triggers, 121, 129, 130
trucks, 53
turtles, 49, 143, 147

U

uncertainty, 8, 52, 62, 69, 70, 76, 87, 95, 96, 109, 110
universe, 2
universities, 119
urban areas, 31
USDA, vii, viii, 2

V

values, 13, 32, 109, 126
variability, 14, 50, 165
variables, 52, 62, 69, 84, 85, 88
vegetation, 73
vessels, 4, 15, 18, 34, 35, 42, 52, 57, 59, 80, 89, 93, 111, 112, 126, 152, 153, 154, 169
veterinarians, 23

W

wastewater, 54
wastewater treatment, 54
water, 27, 34, 40, 55, 62, 66, 72, 80, 84, 86,
  92, 93, 94, 162, 163, 167
wear, 62, 112
welfare, 26, 27
well-being, 23, 28, 54

wholesale, 4, 51
wild animals, 54, 55
wildfires, 73
wildlife, 37, 48
winter, 71, 79, 111
wisdom, 10
World Trade Organization, 39
wrongdoing, 55
WTO, 39, 58